T0226249

This series of short books covers a broad spectrum of titles of interest in electrical engineering that may not specifically fit within another series. Books will focus on fundamentals, methods, and advances of interest to electrical and electronic engineers.

Farzin Asadi

Analog Electronic Circuits Laboratory Manual

 Springer

Farzin Asadi
Department of Electrical and Electronics
Engineering
Maltepe University
Istanbul, Turkey

ISSN 1559-811X   ISSN 1559-8128 (electronic)
Synthesis Lectures on Electrical Engineering
ISBN 978-3-031-25124-5   ISBN 978-3-031-25122-1 (eBook)
https://doi.org/10.1007/978-3-031-25122-1

This Springer imprint is published by the registered company Springer Nature Switzerland AG
The registered company address is: Gewerbestrasse 11, 6330 Cham, Switzerland

Preface

This is a book for a lab course meant to accompany, or follow, any standard course in electronic circuit analysis. It has been written for sophomore or junior electrical and computer engineering students, either concurrently with their electronic circuit analysis class or following that class. This book is appropriate for non-majors, such as students in other branches of engineering and in physics, for which electronic circuits is a required course or elective and for whom a working knowledge of electronic circuits is desirable.

This book has the following objectives:

1. To support, verify, and supplement the theory; to show the relations and differences between theory and practice.
2. To teach measurement techniques.
3. To convince students that what they are taught in their lecture classes is real and useful.
4. To help make students tinkerers and make them used to asking "what if" questions.

This book contains 70 experiments which helps the reader to explore the concepts studied in the classroom. Here is a brief summary of chapters and appendixes:

Chapter 1 focuses on diode circuits. Different types of diodes (i.e., LED, Zener, Schottky), their I-V characteristics, clipping/clamping circuits, voltage doubler circuit, half/full wave rectifier circuits, and switching behavior of diode are studied in this chapter.

Chapter 2 focuses on BJT transistor. I-V characteristics of a BJT transistor, BJT as a switch, switching behavior of BJT, common base, common emitter, and common collector amplifiers are studied in this chapter.

Chapter 3 focuses on MOSFET transistor. I-V characteristics of a MOSFET transistor, MOSFET as a switch, switching behavior of MOSFET, common gate, common source, and common drain amplifiers are studied in this chapter.

Chapter 4 focuses on different types of current sources and differential pairs. Current mirror current source, Widlar current source, Wilson current source, and Op Amp-based current sources are studied in this chapter. You will learn how to measure the common mode gain, differential mode gain, and Common Mode Rejection Ratio (CMRR) for a differential amplifier as well.

Chapter 5 focuses on feedback amplifiers. Voltage series feedback and voltage shunt feedback are studied in this chapter.

Chapter 6 focuses on important applications of Op Amps. Inverting/non-inverting amplifier, buffer, difference amplifier, instrumentation amplifier, window comparator, Schmitt trigger, precious rectifier, filter, integrator, differentiator, and oscillator circuits are studied in this chapter.

Chapter 7 focuses on voltage regulator circuits and linear power amplifiers. Series and shunt voltage regulator, voltage regulator ICs, class B power amplifier (with and without feedback), and class AB power amplifiers are studied in this chapter.

Appendix A shows how to draw different types of graphs with MATLAB®.

Appendix B shows how to design different types of circuits with NI® Multisim™'s Circuit Wizard program.

I hope that this book will be useful to the readers, and I welcome comments on the book.

Istanbul, Turkey   Farzin Asadi
                                                                                      farzinasadi@maltepe.edu.tr

Contents

Diode Circuits

1

1.1 Introduction

A diode is a semiconductor device that essentially acts as a one-way switch for current. It allows current to flow easily in one direction, but severely restricts current from flowing in the opposite direction. Diode is one of the important components which used in electronic circuits. Experiments of this chapter helps you to understand the behavior and applications of this important component.

1.2 Diode Current-voltage (I–V) Characteristic

In this experiment we will study the I–V characteristic of different types of diode. Make the circuit shown in Fig. 1.1. Note that the cathode terminal of diodes has a strip behind it (Fig. 1.2). ID and VD in Fig. 1.1 is a milli ammeter and a voltmeter, respectively.

© The Author(s), under exclusive license to Springer Nature Switzerland AG 2023
F. Asadi, *Analog Electronic Circuits Laboratory Manual*, Synthesis Lectures on Electrical Engineering, https://doi.org/10.1007/978-3-031-25122-1_1

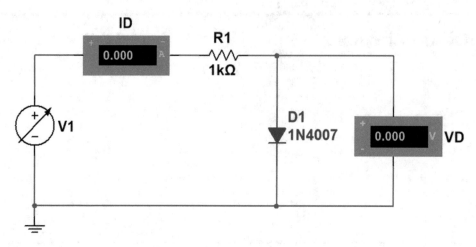

Fig. 1.1   Circuit to measure the I–V characteristic of 1N4007

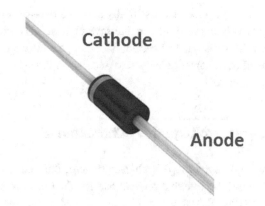

Fig. 1.2   Cathode and anode terminals of the diode

Set the input voltage V1 to the values shown in Table 1.1 and fill the table. Use MATLAB to draw the graph of this data.

Table 1.1   Currents and voltages for 1N4007 diode

V1	−6 V	−4 V	−2 V	0 V	0.5 V	1 V	3 V	5 V	7 V	9 V	11 V	13 V
ID												
VD												

Replace the 1N4007 diode with 1N4148 (Fig. 1.3) and repeat the experiment. Fill the Table 1.2. Use MATLAB to draw the graph of this data.

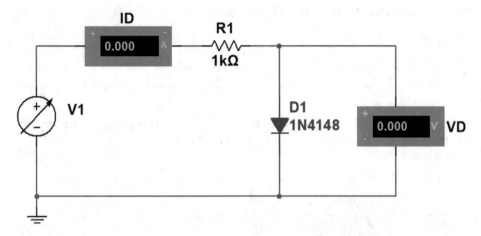

Fig. 1.3 Circuit to measure the I–V characteristic of 1N4148

Table 1.2 Currents and voltages for 1N4148 diode

V1	−6 V	−4 V	−2 V	0 V	0.5 V	1 V	3 V	5 V	7 V	9 V	11 V	13 V
ID												
VD												

Replace the 1N4148 diode with a 3.3 V Zener diode (Fig. 1.4) and repeat the experiment. Fill the Table 1.3. Use MATLAB to draw the graph of this data.

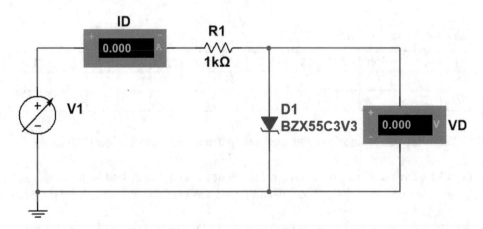

Fig. 1.4 Circuit to measure the I–V characteristic of BZX55C3V3

Table 1.3 Currents and voltages for 3.3 V Zener diode

V1	−6 V	−4 V	−2 V	0 V	0.5 V	1 V	3 V	5 V	7 V	9 V	11 V	13 V
ID												
VD												

Replace the 3.3 V Zener diode with an LED (Fig. 1.5) and repeat the experiment. Fill the Table 1.4. Use MATLAB to draw the graph of this data.

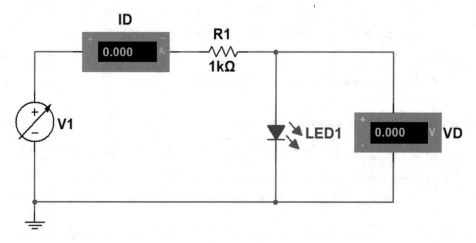

Fig. 1.5 Circuit to measure the I–V characteristic of a LED

Table 1.4 Currents and voltages for LED

V1	−6 V	−4 V	−2 V	0 V	0.5 V	1 V	3 V	5 V	7 V	9 V	11 V	13 V
ID												
VD												

Following points helps you to determine the anode and cathode of an LED easily:

(a) LEDs have one lead that is longer that the other. This longer lead is the anode, and the shorter one is the cathode. Note that we assumed that the leads have not been clipped.

(b) There is a small flat notch on the side of the LED. The lead that is closer to the notch is always the cathode (Fig. 1.6).

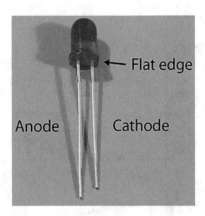

Fig. 1.6 Anode and cathode of a LED

Draw the data of Tables 1.1, 1.2, 1.3 and 1.4 on the same graph and compare the forward voltage drop of studied four diodes with each other. Pay attention to the exponential nature of the I–V graphs in the forward bias region. Which diode has the lowest and which diode has the highest forward voltage drop? Which diode conducts the current in the reverse bias region?

1.3   Test of Diodes with DMM

You can test a diode with DMM and see whether it is good or faulty (i.e. opened or shorted). You can use the DMM to find the anode and cathode terminals of a diode as well.

In order to determine whether a diode is faulty: Put the DMM in the diode test mode and connect the red probe to the anode and the black probe to the cathode (Fig. 1.7). You need to see a forward voltage drop in the range of 0.2–0.8 V for a good diode. The diode is opened if DMM shows OL in this case.

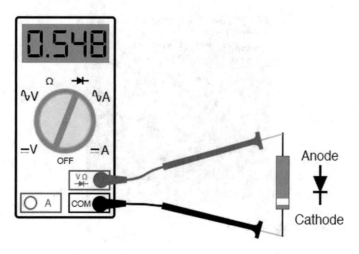

Fig. 1.7 Forward voltage drop of a diode

Now connect the black wire to the anode and red wire to the cathode (Fig. 1.8). You need to see OL for a good diode. If you see anything else the diode is shorted.

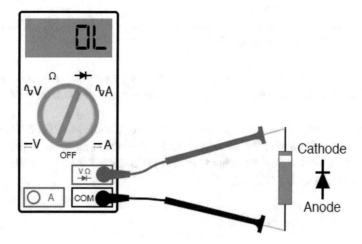

Fig. 1.8 DMM shows OL when red probe and black probe are connected to cathode and anode, respectively

Note that generally an LED cannot be tested with a DMM due to its high forward voltage drop. If you apply the aforementioned procedure to an LED you will observe OL in both cases and by mistake you may think that your LED is faulty (opened). In order to test an LED you need to use a power source to provide enough current for it and see whether it emits light.

Table 1.5 summarize what we said.

Table 1.5 Summary of diode test analysis

Status of the diode under test	Red probe is connected to anode and black probe is connected to cathode	Red probe is connected to cathode and black probe is connected to anode
Good diode	0.2–0.8 V	OL
Opened diode	OL	OL
Shorted diode	Shows a number	Shows a number

Note that if the diode is connected to other components in a circuit (for instance assume that the diode is mounted on a Printed Circuit Board (PCB)) it is necessary to remove at least one end of the diode from the circuit in order to test the diode. Otherwise presence of other components may affect your test and it may lead you to a wrong result

In Figs. 1.2 and 1.6 we learned two features to determine the anode and cathode terminals. Let learn one more feature. In high current diodes with metal body (Fig. 1.9), the metal case is connected to the cathode terminal.

Fig. 1.9 Cathode and anode of a high current metal diode

Here is the general procedure to find out anode and cathode terminals with DMM: Take a DMM and select the diode test mode. Connect two probes to two terminals of the diode, at random. If the DMM shows OL, reverse the probes connections. You must see the forward voltage drop of the diode under test. The terminal connected to red probe is anode and the terminal connected to black probe is cathode.

1.4   Voltage Regulation with Zener Diode

A Zener diode is a heavily doped semiconductor device that is designed to operate in the reverse bias region. When a Zener diode is reveres biased and the potential reaches the Zener voltage (knee voltage), the junction breaks down and the current flows from cathode to anode. This effect is known as the Zener Effect.

Clarence Melvin Zener was the first person to describe the electrical properties of Zener diode. Clarence Zener was a theoretical physicist who worked at Bell Labs. As a result of his work, the Zener diode was named after him. He first postulated the breakdown effect that bears his name in a paper that was published in 1934.

In this experiment we will use a Zener diode to obtain a relatively constant voltage despite of input voltage source changes. Make the circuit shown in Fig. 1.10. ID and VD in Fig. 1.10 is a milli ammeter and a voltmeter, respectively.

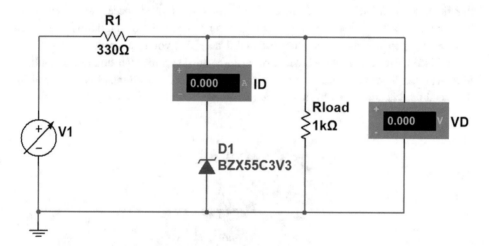

Fig. 1.10 Voltage regulation with Zener diode

Set the input voltage to the values shown in Table 1.6 and fill the table. Note that negative values of V1 in Table 1.6 forward bias the Zener diode. The output voltage is almost constant for big enough positive inputs.

Table 1.6 Currents and voltages for Zener diode

V1	−6 V	−4 V	−2 V	2 V	4 V	6 V	8 V	10 V	12 V
ID									
VD									

Now replace the DC voltage source V1 with a signal generator (Fig. 1.11). Signal generator shown in Fig. 1.11 generates a triangular waveform with peak value of 8 V and frequency of 1 kHz (Fig. 1.12). Set the oscilloscope in the XY mode and observe the output versus input. Draw what you saw on the oscilloscope screen in your lab. notebook carefully and interpret it.

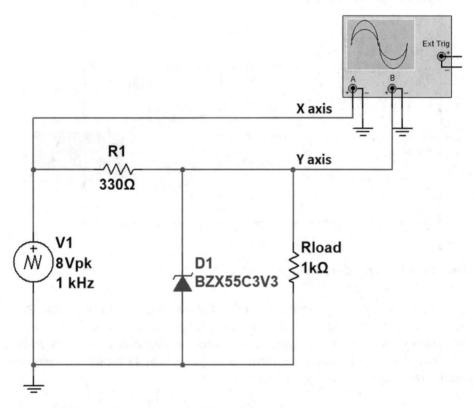

Fig. 1.11   Observing the transfer characteristics of the circuit

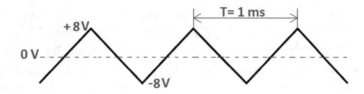

Fig. 1.12   Waveform of voltage source V1

Let's study a simple design problem. We want to design the simple voltage regulator shown in Fig. 1.13. In this figure 10 V < V1 < 15 V and output load (series connection of P1 and R2) changes from 1 kΩ up to 2 kΩ. Value of resistor R2 must be determined in such a way that 3.3 V output is provided for the load despite of input voltage and output load changes. According to the datasheet, minimum value of IZ to keep the Zener diode in the constant voltage region is 5 mA.

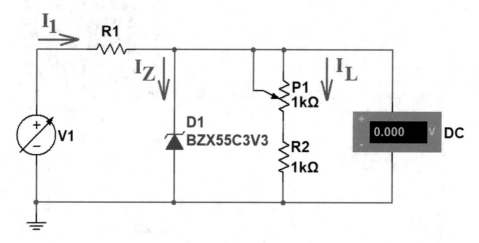

Fig. 1.13 Voltage regulation for a variable load

Table 1.7 shows required values of R1 for different combinations of V1 and P1 + R2. We need to select the minimum value for R1. Therefore, R1 = 0.807 is selected. Selection of minimum value ensures us that always minimum value of 5 mA reaches the Zener diode (test it). Therefore, Zener diode always is in the breakdown region and provide constant voltage of 3.3 V for the load.

Table 1.7 Required value of R1 for different combinations of V1 and P1 + R2

V1 (V)	P1 + R2 (kΩ)	Minimum of I1 (mA)	Value of R1 (kΩ)
10	1	$5 + 3.3/1 = 8.3$	$\frac{10-3.3}{8.3} = 0.807$
10	2	$5 + 3.3/2 = 6.65$	$\frac{10-3.3}{6.65} = 1$
15	1	$5 + 3.3/1 = 8.3$	$\frac{15-3.3}{8.3} = 1.41$
15	2	$5 + 3.3/2 = 6.65$	$\frac{15-3.3}{6.65} = 1.76$

1.5   Clipper Circuit

A clipper is a circuit prevents a signal from exceeding a predetermined reference voltage level. In this experiment we will study the behavior of clipper circuits.

Make the circuit shown in Fig. 1.14. V1 is a triangular waveform with peak value of 8 V and frequency of 50 Hz (Fig. 1.15). Set the oscilloscope to the XY mode in order to observe the circuit transfer function (i.e. the relationship between input and output). Draw what you see on the oscilloscope screen in your lab. notebook and interpret it. Pay attention to clipping level and compare it with value of V2. From theoretical point of view, $V_{load} < 4 + V_d$, where V_d shows the forward voltage drop of the diode. i.e. 0.7 V. Therefore load voltage is limited to $V_{load} < 4 + 0.7 \Rightarrow V_{load} < 4.7$.

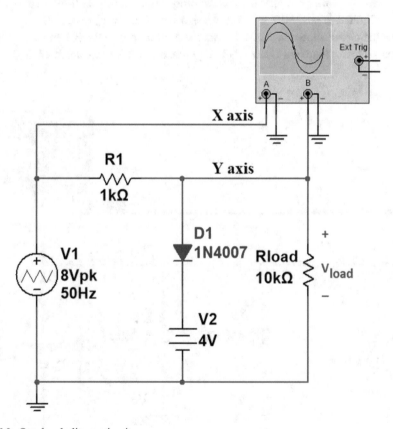

Fig. 1.14 One level clipper circuit

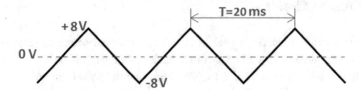

Fig. 1.15 Waveform of voltage source V1

The circuit shown in Fig. 1.14 is a one level clipper. It clips only the positive half cycles of the input and doesn't affect the negative half cycles. The circuit shown in Fig. 1.16 is a two level clipper. This circuit keeps the load voltage (V_{load}) between $-2-V_d$ and $+4+V_d$ Volts, where V_d shows the forward voltage drop of the diode. i.e. 0.7 V. Therefore load voltage is limited to $-2-0.7 < V_{load} < 4+0.7 \Rightarrow -2.7 < V_{load} < 4.7$.

Make the circuit shown in Fig. 1.16. Set the oscilloscope to the XY mode and observe the circuit transfer function. Draw what you observed in your lab. notebook and interpret it.

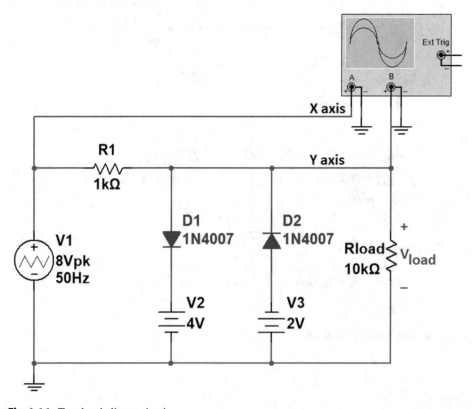

Fig. 1.16 Two level clipper circuit

1.6   Clamping Circuit

A Clamper Circuit is a circuit that adds a DC level to an AC signal. Actually, the positive and negative peaks of the signals can be placed at desired levels using the clamping circuits. As the DC level gets shifted, a clamper circuit is called as a Level Shifter. In this experiment clamp circuits are studied.

Make the circuit shown in Fig. 1.17. Draw the waveforms that you see on the scope carefully in your lab. notebook. Interpret what you see on the oscilloscope screen.

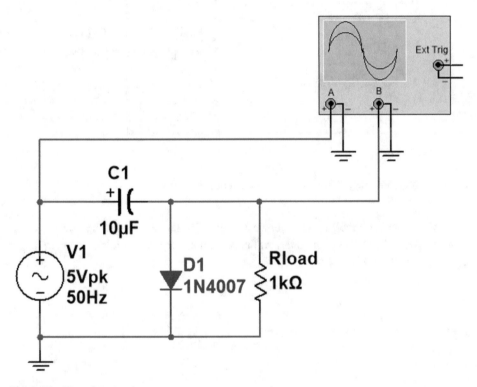

Fig. 1.17 Clamping circuit

Connect a DC voltmeter to the capacitor C1 and measure its voltage (Fig. 1.18). What is the relationship between capacitor voltage and peak value of input voltage source V1?

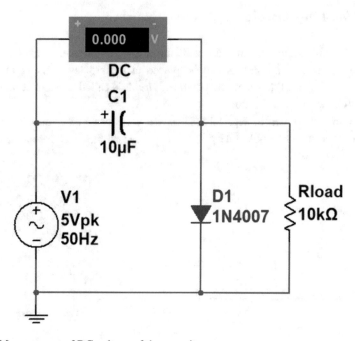

Fig. 1.18 Measurement of DC voltage of the capacitor

Now change the circuit to what shown in Fig. 1.19. Draw the input and output wave-forms in your lab. notebook and interpret them. Compare the output waveform with output of circuit shown in Fig. 1.17.

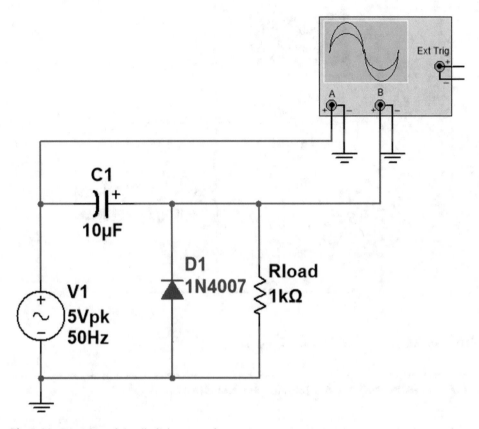

Fig. 1.19 Direction of the diode is reversed

Connect a DC voltmeter to the capacitor C1 and measure its voltage (Fig. 1.20). What is the relationship between capacitor voltage and peak value of input voltage source V1?

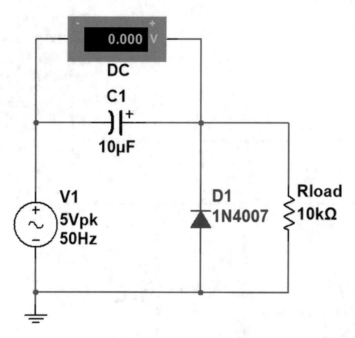

Fig. 1.20 Measurement of DC voltage of the capacitor

1.7   Small Signal AC (Dynamic) Resistance of Diode

The current–voltage relationship for a diode is given by $I_D = I_s \left(e^{\frac{V_D}{nV_T}} - 1 \right)$. I_D and V_D show the diode current and diode voltage ($V_{anode} - V_{cathode}$), respectively. I_s is the reverse saturation current and n is the emission coefficient. The emission coefficient n depends on the fabrication process and semiconductor material (generally $1 < n < 2$). V_T is the thermal voltage and it is 25.85 mV for 300 K. The AC (dynamic) resistance of the diode is defined as $r_d = \frac{dV_D}{dI_D} \approx \frac{nV_T}{I_D}$.

In this experiment we want to measure the AC resistance of a diode. Make the circuit shown in Fig. 1.21 Note that there is no need to use a signal generator and a DC source. Required 3 V can be generated with the DC offset knob of signal generator. RO shows the output resistance of the signal generator. Select the AC coupling for two channels. This permits you to observe only the AC component on the oscilloscope screen. Measure the peak value of channel A and B call them $V_{A,p}$ and $V_{B,p}$, respectively.

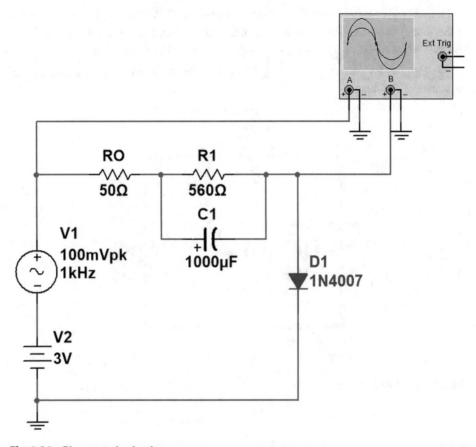

Fig. 1.21 Given sample circuit

The DC equivalent of Fig. 1.21 is shown in Fig. 1.22. The DC current of diode is around $\frac{3-0.7}{50+560} = 3.77\,\text{mA}$.

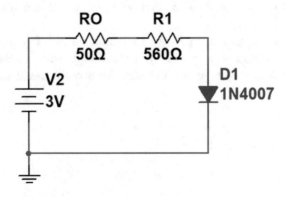

Fig. 1.22 Equivalent DC circuit

The AC equivalent of Fig. 1.21 is shown in Fig. 1.23 (RD shows the AC resistance of the diode). Note that the capacitor acts like a short circuit since its impedance is very low at 1 kHz: $\frac{-j}{2\pi fC} = \frac{-j}{2\pi \times 1k \times 100\mu} = -j0.159$. The AC resistance of diode equals to $\frac{R_D}{R_O+R_D}V_{A,p} = V_{B,p} \Rightarrow R_D = \frac{V_{B,p}}{V_{A,p}-V_{B,p}}R_O$. Use this relationship to calculate the AC resistance of the diode.

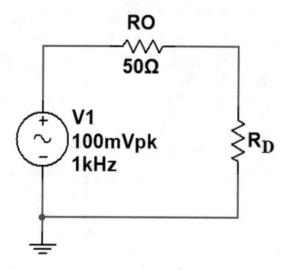

Fig. 1.23 Equivalent AC circuit

1.8 Voltage Doubler

A voltage doubler is a circuit which charges capacitors from the input voltage and switches these charges in such a way that, in the ideal case, exactly twice the voltage is produced at the output as at its input. In this circuit we will study the voltage doubler circuit.

Make the circuit shown in Fig. 1.24 and measure the load voltage (capacitor C2 voltage). Compare the measured value with the peak of input voltage source V1. Is it two times the peak of input voltage source V1? Describe how the circuit shown in Fig. 1.24 works.

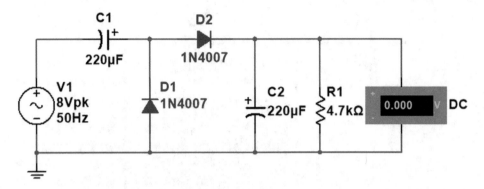

Fig. 1.24 Voltage doubler circuit

1.9   Half-Wave Rectifier

A half-wave rectifier is the simplest form of the rectifier and requires only one diode for the construction. In this experiment we will study the half wave rectifiers.

Make the circuit shown in Fig. 1.25. T1 is a small step down transformer. For instance, you can use a transformer with output voltage of 6 Vrms.

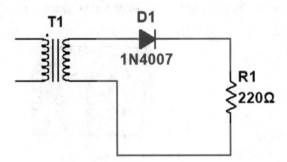

Fig. 1.25 Half wave rectifier circuit

Use the oscilloscope to observer the secondary winding's waveform (Fig. 1.26). Measure the peak value and frequency of observed voltage and call it $V_{p,in}$ and f, respectively.

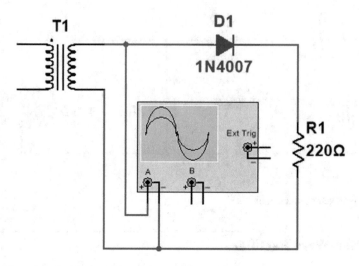

Fig. 1.26 Observing the input voltage of the rectifier

Connect the oscilloscope to the diode D1 (Fig. 1.27) and observe the diode's voltage waveform. Interpret what you see on the oscilloscope screen. Determine the forward bias and reverse bias regions for the waveform you see on the oscilloscope screen. Measure the maximum reverse voltage applied to the diode and compare it with $V_{p,in}$ as well.

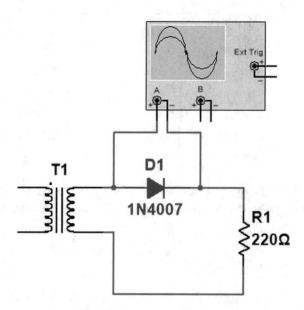

Fig. 1.27 Observing the voltage of the diode

Observe the input and output simultaneously (Fig. 1.28). Compare the peak value of load voltage with $V_{p,in}$.

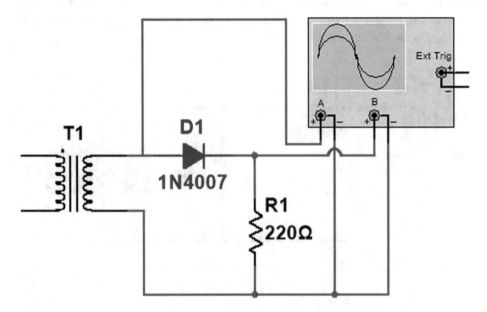

Fig. 1.28 Observing the input and output simultaneously

Use a DC voltmeter to measure the average value (DC component) of load voltage (Fig. 1.29). Compare the measured value with the value with value predicted by theory, i.e., $\frac{V_{p,in}}{\pi}$. Note that forward voltage drop of the diode is ignored in extraction of $\frac{V_{p,in}}{\pi}$ formula. Therefore, some discrepancy between measured value and theoretical value is expected.

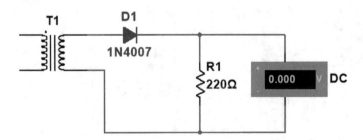

Fig. 1.29 Measurement of average value of output

Use a true RMS AC voltmeter (Fig. 1.30) or a digital oscilloscope (Fig. 1.31) to measure the RMS value of load voltage (Use the Measurement menu of the oscilloscope).

Compare the measured value with value predicted by theory, i.e., $\frac{V_{p,in}}{2}$. Note that forward voltage drop of the diode is ignored in extraction of $\frac{V_{p,in}}{2}$ formula. Therefore, some discrepancy between measured value and theoretical value is expected. Average power dissipated in resistor R1 can be calculated with the aid of $P_{R1,ave} = \frac{V_{R1,RMS}^2}{R1}$ where $V_{R1,RMS}$ is the RMS of resistor $R1$'s voltage.

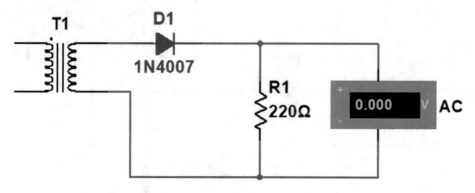

Fig. 1.30 Measurement of RMS value of output with an AC voltmeter

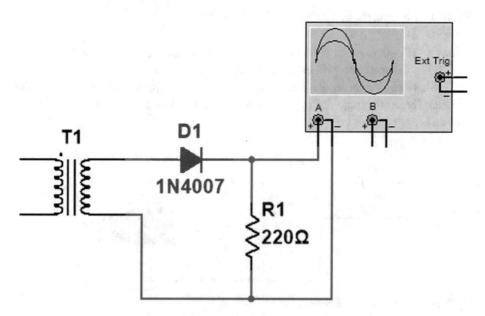

Fig. 1.31 Measurement of RMS value of output with an oscilloscope

Now add a filter capacitor to the circuit (Fig. 1.32). The DC voltmeter in Fig. 1.32 measures the average value (DC component) of load voltage. Coupling of the oscilloscope in Fig. 1.32 is set to AC and it measures the peak-to-peak and frequency of load's ripple voltage.

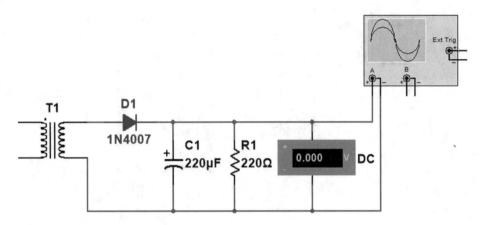

Fig. 1.32 Half wave rectifier with filter capacitor

Fill the Table 1.8 for given values of capacitor C1. In this table, Vr(p − p), Fr and VR1(DC) show peak-to-peak of load's ripple voltage, frequency of load's ripple voltage and average value (DC component) of load's voltage, respectively.

Table 1.8 Measured values for circuit shown in Fig. 1.32

C1 (μF)	Vr(p − p)	Fr	VR1(DC)
220			
470			

Compare the measured values with values predicted by theory ($Vr(p − p) = \frac{V_{p,in}}{f \times R1 \times C1}$, $Fr = 2f$ and $VR1(DC) = V_{p,in}$). Note that forward voltage drop of the diode is ignored in extraction of $Vr(p − p) = \frac{V_{p,in}}{f \times R1 \times C1}$ and $VR1(DC) = V_{p,in}$ formulas. Therefore, some discrepancy between measured value and theoretical value is expected.

1.10 Full Wave Rectifier

In this experiment we will study the full-wave rectifier. Make the circuit shown in Fig. 1.33. You can make a full wave rectifier with a center tap transformer and two diodes as well (Fig. 1.34).

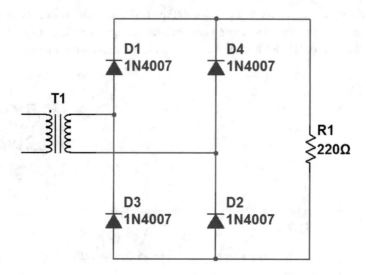

Fig. 1.33 Full wave rectifier

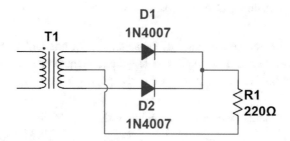

Fig. 1.34 Full wave rectifier with center tap transformer

Let's continue our experiment with respect to the rectifier shown in Fig. 1.33. Use the oscilloscope to observer the secondary winding's waveform (Fig. 1.35). Measure the peak value and frequency of observed voltage and call it $V_{p,in}$ and f, respectively.

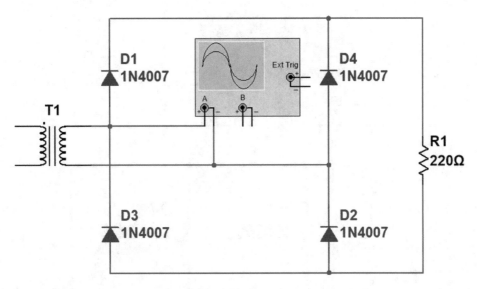

Fig. 1.35 Observing the input voltage to the rectifier

Connect the oscilloscope to the diode D1 (Fig. 1.36) and observe the diode's voltage waveform. Interpret what you see on the oscilloscope screen. Determine the forward bias and reverse bias regions for the waveform you see on the oscilloscope screen. Measure the maximum reverse voltage applied to the diode and compare it with $V_{p,in}$ as well.

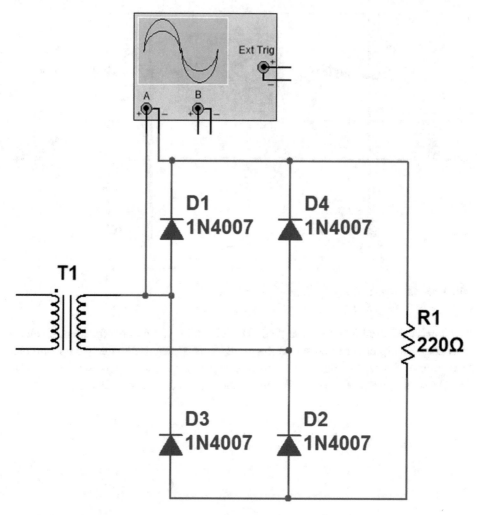

Fig. 1.36 Observing the diode's voltage waveform

Observe the output voltage waveform (Fig. 1.37) and measure its frequency and peak value. Compare the measured frequency and peak value with f and $V_{p,in}$, respectively.

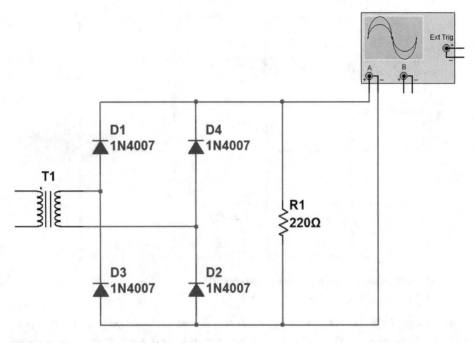

Fig. 1.37 Observing the output

Use a DC voltmeter to measure the average value (DC component) of load voltage (Fig. 1.38). Compare the measured value with the value with value predicted by theory, i.e., $\frac{2 \times V_{p,in}}{\pi}$. Note that forward voltage drop of the diode is ignored in extraction of $\frac{2 \times V_{p,in}}{\pi}$ formula. Therefore, some discrepancy between measured value and theoretical value is expected.

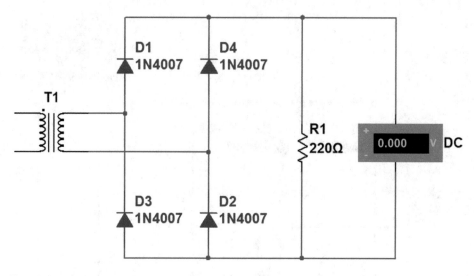

Fig. 1.38 Measurement of average value of the output

Use a true RMS AC voltmeter (Fig. 1.39) or a digital oscilloscope (Fig. 1.40) to measure the RMS value of load voltage. Compare the measured value with value predicted by theory, i.e., $\frac{V_{p,in}}{\sqrt{2}}$. Note that forward voltage drop of the diode is ignored in extraction of $\frac{V_{p,in}}{\sqrt{2}}$ formula. Therefore, some discrepancy between measured value and theoretical value is expected. Average power dissipated in resistor R1 can be calculated with the aid of $P_{R1,ave} = \frac{V_{R1,RMS}^2}{R1}$ where $V_{R1,RMS}$ is the RMS of resistor R1's voltage.

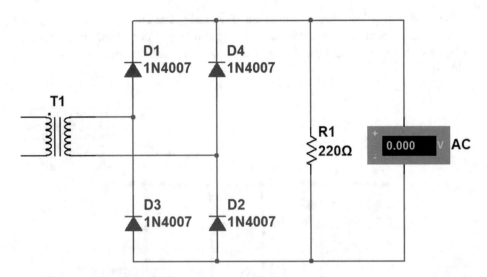

Fig. 1.39   Measurement of RMS value of output with an AC voltmeter

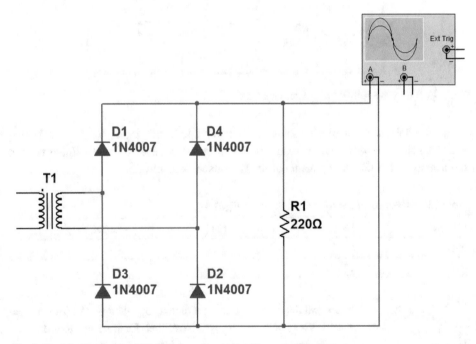

Fig. 1.40   Measurement of RMS value of output with an oscilloscope

Now add a filter capacitor to the circuit (Fig. 1.41). The DC voltmeter in Fig. 1.41 measures the average value (DC component) of load voltage. Coupling of the oscilloscope in Fig. 1.41 is set to AC and it measures the peak-to-peak and frequency of load's ripple voltage.

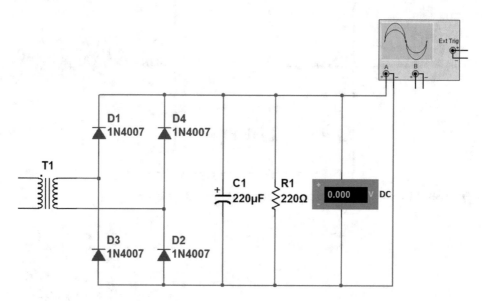

Fig. 1.41 Full wave rectifier with filter capacitor

Fill the Table 1.9 for given values of capacitor C1. In this table, Vr(p − p), Fr and VR1(DC) show peak-to-peak of load's ripple voltage, frequency of load's ripple voltage and average value (DC component) of load's voltage, respectively.

Table 1.9 Measured values for circuit shown in Fig. 1.41

C1 (μF)	Vr(p − p)	Fr	VR1(DC)
220			
470			

Compare the measured values with values predicted by theory $(Vr(p - p) = \frac{V_{p,in}}{2 \times f \times R1 \times C1}$, $Fr = 2f$ and $VR1(DC) = V_{p,in})$. Note that forward voltage drop of the diodes is ignored in extraction of $Vr(p - p) = \frac{V_{p,in}}{2 \times f \times R1 \times C1}$ and $VR1(DC) = V_{p,in}$ formulas. Therefore, some discrepancy between measured value and theoretical value is expected.

1.11 Full-Wave Rectifier with Zener Diode Regulator

The full-wave rectifier can be combined with voltage regulator circuits in order to provide a constant voltage for sensitive loads like digital circuits. In this experiment a full-wave rectifier is combined with a simple Zener regulator to provide 3.3 V for the load.

Make the circuit shown in Fig. 1.42. The transformer T1 in Fig. 1.42 provides a 6 Vrms for the rectifier. Measure the load voltage with DC voltmeter and ensure that it provides 3.3 V for the load. Explain how this circuit works.

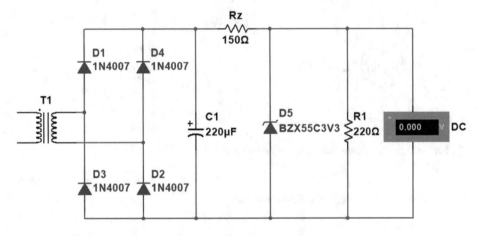

Fig. 1.42 Full-wave rectifier with Zener diode regulator

1.12 Switching Behavior of Diode

In this experiment switching times of diode is studied. You will see that a diode can't go from forward bias to reverse bias instantaneously. Let's get started. Make the circuit shown in Fig. 1.43 (Coupling of both channel is set to DC). RO shows the output resistance of the signal generator. Signal generator V1 generates the waveform shown in Fig. 1.44. Frequency of this pulse is 50 kHz.

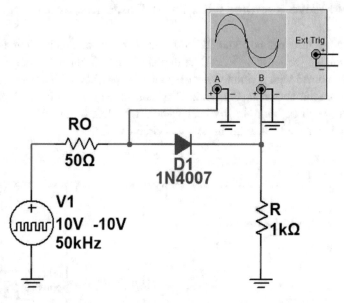

Fig. 1.43   Circuit to test the switching speed of the diode

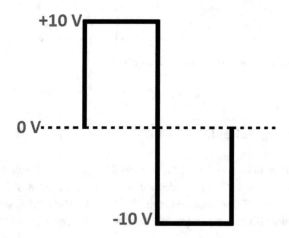

Fig. 1.44   Waveform of voltage source V1

Observe and draw the output voltage waveform in your lab. notebook. Pay attention to the region that diode goes from forward bias to reverse bias and measure the storage t_s and transition time t_t (Fig. 1.45). Summation of t_s and t_t givs the reverse recovery time t_{rr}.

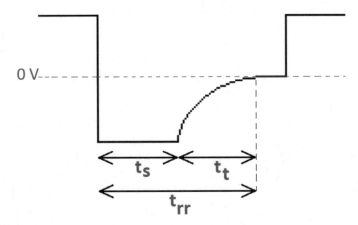

Fig. 1.45 Switching behavior of the diode

Replace the 1N4007 diode with 1N4148 and repeat the experiment. Compare the switching behavior of this two diodes with each other.

References for Further Study

1. Asadi F., Essential Circuit Analysis using Proteus, Springer, 2022. DOI: https://doi.org/10.1007/978-981-19-4353-9
2. Asadi F., Essential Circuit Analysis using LTspice, Springer, 2022. DOI: https://doi.org/10.1007/978-3-031-09853-6
3. Asadi F., Essential Circuit Analysis using NI Multisim and MATLAB, Springer, 2022. DOI: https://doi.org/10.1007/978-3-030-89850-2
4. Asadi F., Electric Circuit Analysis with EasyEDA, Springer, 2022. DOI: https://doi.org/10.1007/978-3-031-00292-2

Transistor Amplifiers and Their Frequency Response

2

2.1 Introduction

Transistors are one of the important components that are used in electronic circuits. Transistors are used for amplification or switching purposes. This chapter focuses on Bipolar Junction Transistors (BJT). Most of the circuits in this chapter use 2N3904 transistor. Pins of 2N3904 transistor are shown in Fig. 2.1. Pinout of a transistor can be found in its datasheet or Google Images.

Fig. 2.1 Pinout of 2N3904

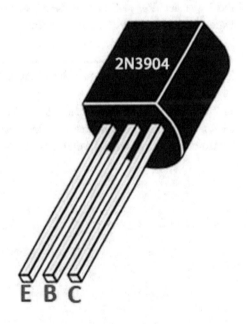

2.2   Testing a Transistor with DMM

This experiment shows how you can test a transistor with a DMM. Use the transistor datasheet to determine its pinout. Then,

(A) In order to test an NPN transistor: (1) Put the DMM in the diode test mode, connect the red probe to the base and black probe to the emitter. The number shown on the dispaly is the voltage drop of base-emitter junction. Let's call the shown value with V_{BEJ}. Now connect the black wire to collector (red is connected to base) and measure the voltage drop of base-collector junction. Let's call the shown value with V_{BCJ}. For a good transistor: (1) Both of V_{BEJ} and V_{BCJ} are between 0.5 and 0.8 V. (2) V_{BEJ} is a little bit bigger than V_{BCJ}.

(B) In order to test an NPN transistor: (1) Put the DMM in the diode test mode, connect the black probe to the base and red probe to the emitter. The number shown on the dispaly is the voltage drop of base-emitter junction. Let's call the shown value with V_{BEJ}. Now connect the red wire to collector (black is connected to base) and measure the voltage drop of base-collector junction. Let's call the shown value with V_{BCJ}. For a good transistor: (1) Both of V_{BEJ} and V_{BCJ} are between 0.5 and 0.8 V. (2) V_{BEJ} is a little bit bigger than V_{BCJ}.

Let's get started. Apply the aforementioned the 2N3904 NPN transistor and measure the V_{BEJ} and V_{BCJ}. Compare the V_{BEJ} with V_{BCJ}. Which one is bigger? Repeat the experiment with 2N3906 PNP transistor.

2.3   Current-Voltage Characteristic (I–V) of Transistor

In this experiment we will extract the I–V characteristic of a transistor. Make the circuit shown in Fig. 2.2. VBB controls the base current. The base current can be calculated by measurement of resistor R1 voltage (i.e., VR1) and using the Ohm's law ($I_B = \frac{V_{R1}}{R1}$).

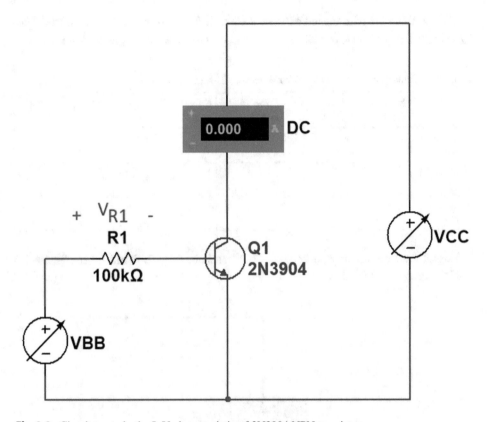

Fig. 2.2   Circuit to study the I–V characteristic of 2N3904 NPN transistor

Set the VBB to the values shown in Table 2.1 and fill it. Draw the graph of this table (i.e., collector current vs. collector-emitter voltage) with MATLAB.

Table 2.1   Collector current for different VBB and VCC

	Base current (IB = VR1/R1)	Collector current for VCC = 0.5 V	Collector current for VCC = 2 V	Collector current for VCC = 5 V	Collector current for VCC = 10 V	Collector current for VCC = 12 V	Collector current for VCC = 15 V
VBB = 2.7 V							
VBB = 4.7 V							
VBB = 6.7 V							

Fill the Table 2.2 with the aid of Table 2.1's data. The current gain β can be calculated by using the $\frac{I_C}{I_B}$ formula. I_C and I_B show the collector and base currents, respectively.

Table 2.2 Transistor current gain (β) for different VBB and VCC

	β for VCC = 0.5 V	β for VCC = 2 V	β for VCC = 5 V	β for VCC = 10 V	β for VCC = 12 V	β for VCC = 15 V
VBB = 2.7 V						
VBB = 4.7 V						
VBB = 6.7 V						

Repeat the experiment with 2N2904 PNP transistor (Fig. 2.3) and fill the Tables 2.3 and 2.4. Pinout of 2N2904 is shown in Fig. 2.4.

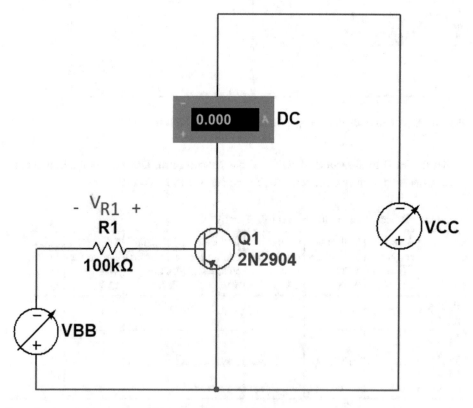

Fig. 2.3 Circuit to study the I–V characteristic of 2N2904

Table 2.3   Collector current for different VBB and VCC

	Base current (IB = VR1/R1)	Collector current for VCC = 0.5 V	Collector current for VCC = 2 V	Collector current for VCC = 5 V	Collector current for VCC = 10 V	Collector current for VCC = 12 V	Collector current for VCC = 15 V
VBB = 2.7 V							
VBB = 4.7 V							
VBB = 6.7 V							

Table 2.4   Transistor current gain (β) for different VBB and VCC

	β for VCC = 0.5 V	β for VCC = 2 V	β for VCC = 5 V	β for VCC = 10 V	β for VCC = 12 V	β for VCC = 15 V
VBB = 2.7 V						
VBB = 4.7 V						
VBB = 6.7 V						

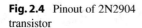

Fig. 2.4 Pinout of 2N2904
transistor

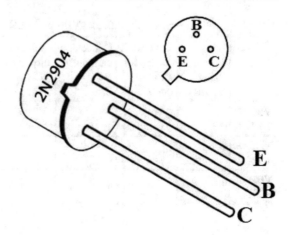

2.4   Transistor as a Switch

In this experiment we will use the transistor as a switch. We want to drive a small 12 V
relay. The relay coil can be modeled as an RL load. When the relay is connected to 12 V,
steady state current of 16.6 mA is drawn from the source, therefore the resistance of the
relay is R = 722 Ω. Value of inductance is not important in this section, but we need to
be aware that the load (relay's coil) is inductive.

Assume a design similar to the one shown in Fig. 2.5. When S1 is connected to ground
the transistor is in the cut-off and no current pass through the collector and the relay is
not activated.

When S1 is connected to +5 V, base current of $I_B = \frac{5-0.7}{12} = 358\,\mu A$ is passed from
the transistor. The transistor has minimum current gain of $\beta_{min} = 50$. Therefore, the
transistor enters the saturation and collector current of 16.6 mA pass through the relay
and activates it.

Assume that transistor is saturated and 16.6 mA pass through the relay. Now you apply
0 V to the base. This forces the transistor to go into the cut-off region and the collector
current needs to reach zero in a short time. This rapid current change and presence of
inductor may generate a big voltage across the collector emitter terminals and may damage
the transistor and circuit shown in Fig. 2.5 is not recommended for RL loads.

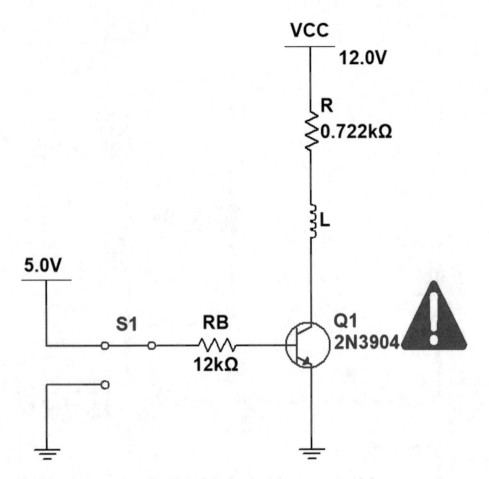

Fig. 2.5 Transistor switch with RL load. Such a circuit is not recommended

A diode must be added to protect the transistor if the transistor load is RL (Fig. 2.6). Diode D1 in Fig. 2.6 is called free wheeling diode.

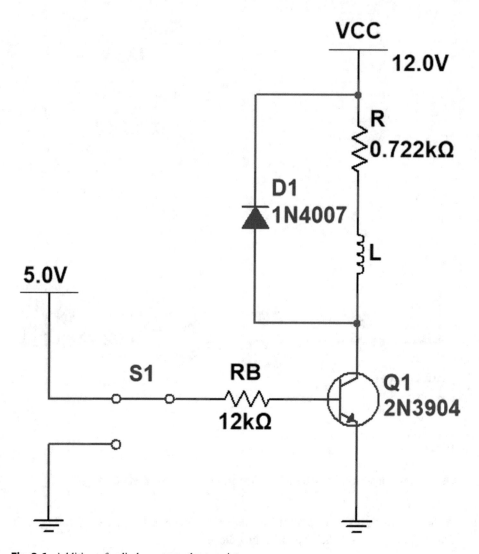

Fig. 2.6 Addition of a diode protects the transistor

Let's see how the diode D1 in Fig. 2.6 protects the transistor. When 5 V is applied to the base, the transistor is saturated and the diode is reverse biased. Diode D1 has no role to play during this phase and all the load current pass through the transistor (Fig. 2.7).

When 0 V is applied to the base, the transistor goes into the cut-off region, the diode is forward biased and the load current pass through the diode D1 (Fig. 2.8). Transistor Q1 has no role to play during this phase. The energy stored in the inductor dissipates in the resistor R and diode D1 and doesn't destroy the transistor. Collector emitter voltage is equal to VCC during this phase.

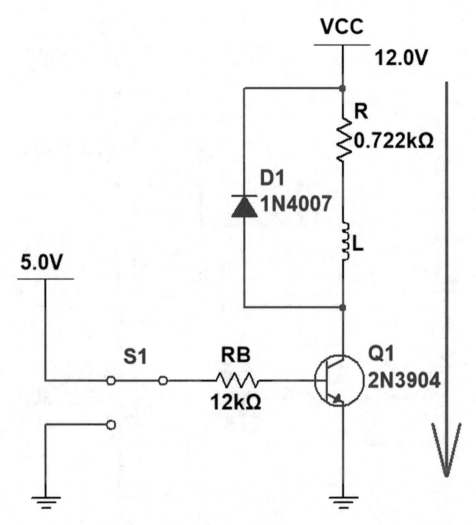

Fig. 2.7 Conduction path for the case that transistor is closed

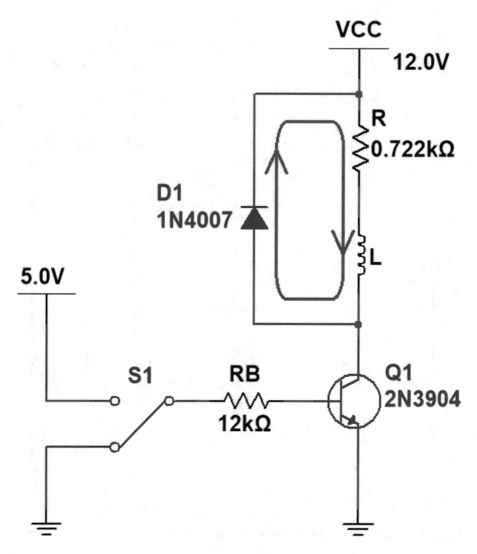

Fig. 2.8 Conduction path for the case that transistor is opened

Now we are ready to start the experiment. Consider the circuit shown in Fig. 2.9. The R and L in Fig. 2.9 represents the 12 V relay's coil resistance and inductance, respectively. Measure the resistance of the coil (R) with a DMM. The required current for the relay is $I_C = \frac{V_{CC}}{R} = \frac{12}{R}$. Let's take the minimum current gain (β_{min}) to be 50 (you can take even lower values, e.g., 30, if you like to be conservative). Therefore, the required base current is $I_B = \frac{I_C}{\beta_{min}} = \frac{\frac{V_{CC}}{R}}{\beta_{min}} = \frac{V_{CC}}{R \times \beta_{min}}$. This means, $I_B = \frac{V_{CC}}{R \times \beta_{min}} = \frac{V_{CC} - V_{BE}}{R_B} = \frac{V_{CC} - 0.7}{R_B}$ or $R_B = \frac{(V_{CC} - 0.7) \times R \times \beta_{min}}{V_{CC}} = 47 \times R$. Select a standard value close to $47 \times R$ for R_B and make the circuit. Use the switch S1 to open/close the relay several times. The transistor will not be destroyed and continue working since it is protected by the diode D1.

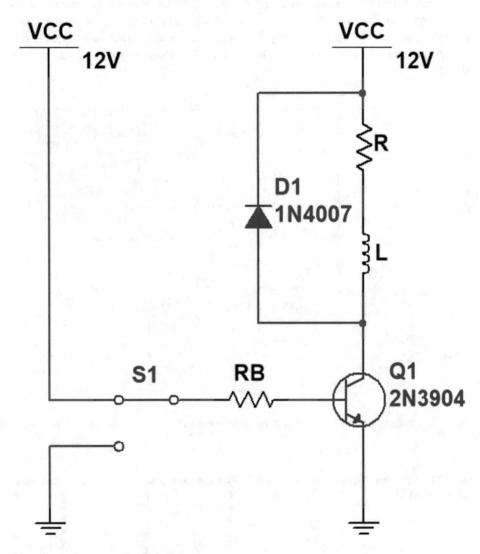

Fig. 2.9 Transistor switch with RL load

2.5   Switching Times of Transistor

In this experiment switching times of a transistor is studied. You will see that a transistor can't go from saturation to cut-off instantaneously. Let's get started. Make the circuit shown in Fig. 2.10. RO shows the output resistance of the signal generator. Signal generator V1 generates the waveform shown in Fig. 2.11. Frequency of generated waveform is 50 kHz.

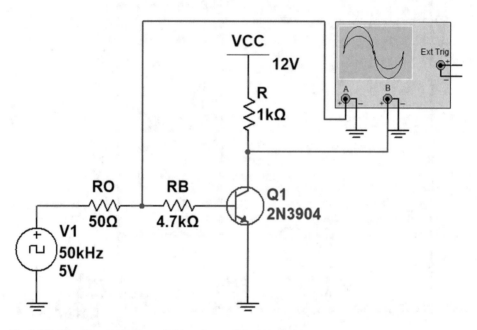

Fig. 2.10 Circuit to study the switching times of the transistor

Fig. 2.11 Waveform of input voltage source V1

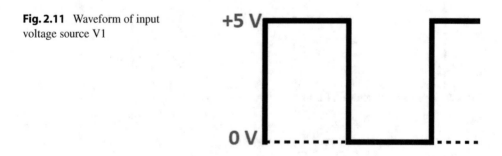

Draw what you see on the oscilloscope screen in your notebook. Measure the t_{on} and t_{off} (Fig. 2.12).

Fig. 2.12 Switching times of
transistor

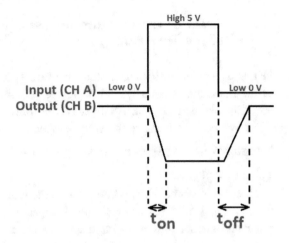

Now add a 100 pF in parallel with RB (Fig. 2.13). Measure the t_{on} and t_{off} again. Explain the role of capacitor C1.

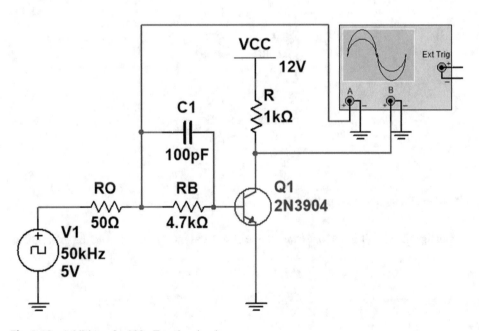

Fig. 2.13   Addition of a 100 pF to the circuit

2.6   Effect of Temperature on the Operating Point of the Transistor

In this experiment we want to study the effect of temperature on the operating point of the transistor. A good bias must not be sensitive to temperature and parameters of transistor like current gain. Let's get started. Make the circuit shown in Fig. 2.14. Use a DC voltmeter to measure the voltage of nodes. Determine the operating point of the transistor ($V_{CE} = V_C - V_E$ and $I_C = \frac{V_{CC} - V_C}{RC}$) and its current gain ($\beta = \frac{I_C}{I_B} = \frac{\frac{V_{CC}-V_C}{RC}}{\frac{V_{CC}-V_B}{RB}} = \frac{RB}{RC} \times \frac{V_{CC}-V_C}{V_{CC}-V_B}$).

Now, keep a hot soldering iron close to the transistor for 15 s. After 15 s re-measure the DC voltage of nodes and calculate the operating point of the transistor. Compare the calculated number with the before heating values and calculate the percentage of change.

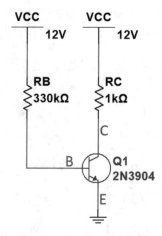

Fig. 2.14 Biasing with base resistor

Change the circuit to the ones shown in Figs. 2.15 and 2.16 and repeat the experiment.

Fig. 2.15 Voltage divider bias
without emitter resistance

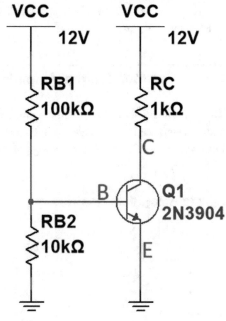

Fig. 2.16 Voltage divider bias
with emitter resistance

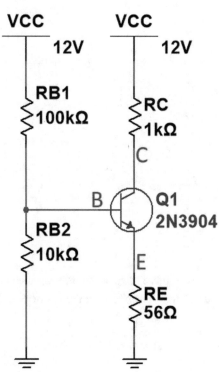

Which one of the circuits has a smaller operating point changes in response to the temperature change?

2.7   Common Emitter Amplifier

This experiment studies the common emitter amplifier. Make the circuit shown in Fig. 2.17 and measure the base voltage (V_C), collector voltage (V_C) and emitter voltage (V_E). Use these values to calculate the operating point of the transistor ($I_{CQ} = \frac{V_{CC}-V_C}{R_3}$ and $V_{CEQ} = V_C - V_E$).

Fig. 2.17 Voltage divider bias

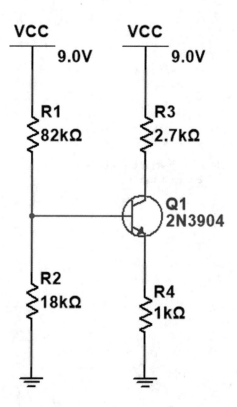

Use a micro ammeter to measure the base current I_{BQ} (Fig. 2.18). Calculate the current gain of the transistor with the aid of $\beta = \frac{I_{CQ}}{I_{BQ}} = \frac{V_{CC}-V_C}{R_3 \times I_{BQ}}$.

Fig. 2.18 Measurement of base current

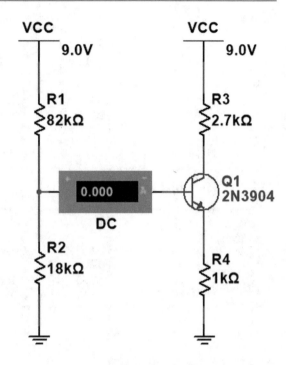

Change the circuit to what shown in Fig. 2.19 (set the coupling of both channels to AC). Rs represents the output impedance of the signal generator Vs. Signal generator Vs generates a sinusoidal signal with peak value of 10 mV and frequency of 1 kHz.

Measure the voltage gain of the amplifier shown in Fig. 2.19. Note that 180° of phase difference exists between input and output (voltage of load resistor R5) and output waveform is distorted, i.e., positive and negative half cycles of input signal are amplified with different gains. Let's simply define the voltage gain as ratio of peak-to-peak of output voltage to peak-to-peak of input voltage.

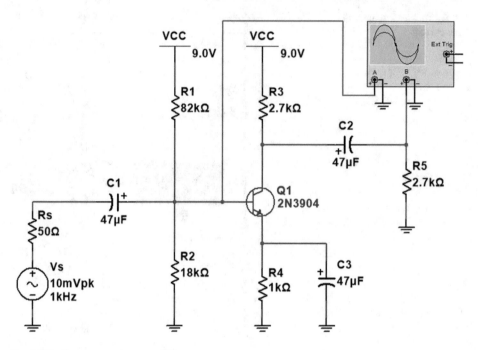

Fig. 2.19 Common emitter amplifier

Obtaining a 10 mV output voltage with old signal generators may be little bit difficult. If you are using an old analog signal generator, then you can use the circuit shown in Fig. 2.20. In this circuit signal generator produces 1 V and a voltage divider helps to obtain around 10 mV. Thevenin's equivalent of this circuit is shown in Fig. 2.21. It is quite close to what we need in Fig. 2.19. Therefore, Fig. 2.22 is equal to Fig. 2.19. You can use this technique in other experiments as well.

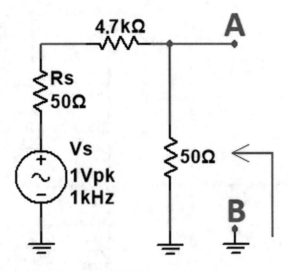

Fig. 2.20 Generation of 10 mVpk from a 1 Vpk source

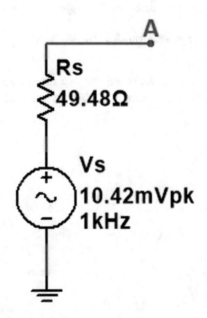

Fig. 2.21 Thevenin equivalent circuit for points A and B

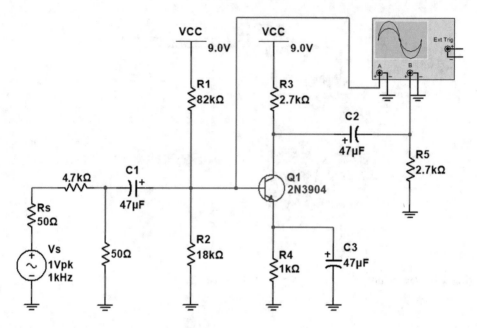

Fig. 2.22 Common emitter amplifier is supplied with the circuit shown in Fig. 2.20

Remove the capacitor C3 (Fig. 2.23) and measure the voltage gain. Compare the voltage gain of this circuit with the one shown in Figs. 2.19 or 2.22. What is the effect of R4 on the voltage gain?

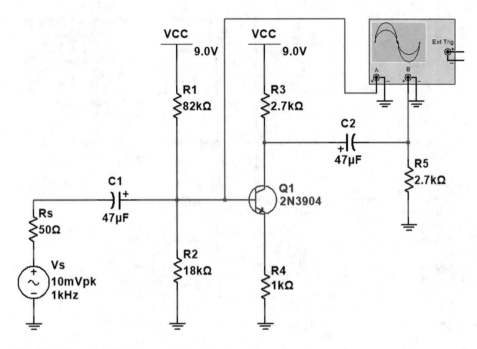

Fig. 2.23 Capacitor C3 is removed from the circuit

Now connect the channel B to collector terminal (Fig. 2.24) and set the coupling of channel B to DC. Increase the amplitude of input and measure the maximum output swing without any clipping.

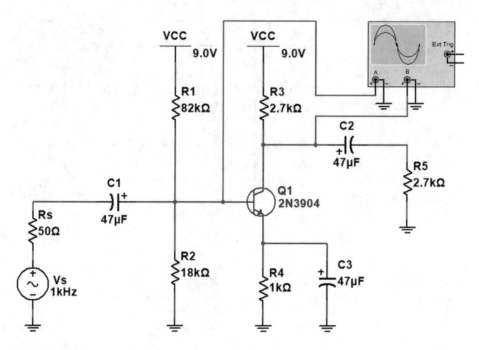

Fig. 2.24 Measurement of maximum output swing

Do a hand analysis for the circuits shown in Figs. 2.19 and 2.23 calculate the voltage gain (use the current gain that you calculated in the beginning of this experiment, i.e., $\beta = \frac{I_{CQ}}{I_{BQ}} = \frac{V_{CC}-V_C}{R_3 \times I_{BQ}}$). Compare the calculated values with the measured ones.

2.8   Measurement of Input Resistance for Common Emitter Amplifier

In this experiment we want to measure the input resistance (Fig. 2.25) of a common emitter amplifier. The technique studied in this experiment can be used to measure the input resistance of other type of amplifiers as well.

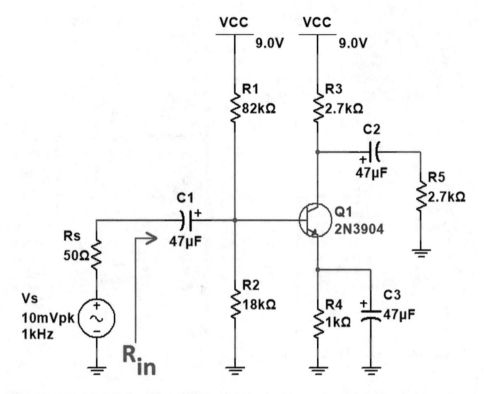

Fig. 2.25 Input resistance of the amplifier

Open the connection between the input signal source and the amplifier and measure the peak-to-peak of input signal (Fig. 2.26). Let's call this value open circuit peak-to-peak $V_{oc,p-p}$.

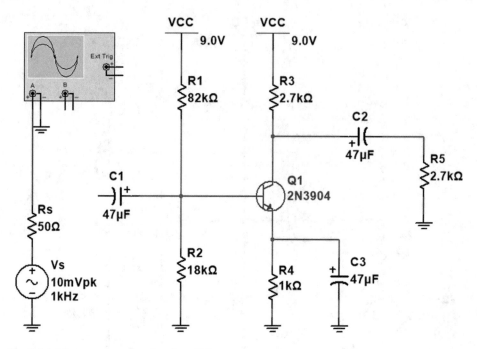

Fig. 2.26 Measurement of peak value of Vs

Put a 1 kΩ resistor Rx in series with the signal generator and measure the peak-to-peak of node X and call it loaded peak-to-peak $V_{loaded,p-p}$ (Fig. 2.27).

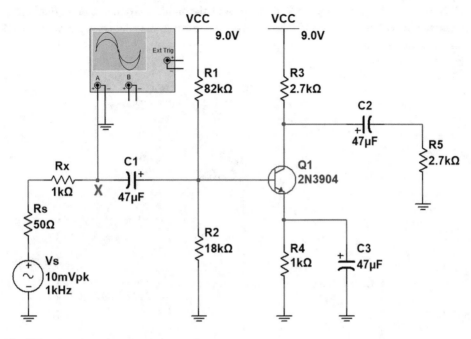

Fig. 2.27   Measurement of peak value of node X

Equivalent circuit for Fig. 2.27 is shown in Fig. 2.28. Input resistance R_{in} (in kΩ) can be calculated using $\frac{R_{in}}{R_{in}+R_x+R_s} V_{oc,p-p} = V_{loaded,p-p} \Rightarrow \frac{R_{in}}{R_{in}+1+0.05} V_{oc,p-p} = V_{loaded,p-p} \Rightarrow R_{in} = \frac{1.05}{\frac{V_{oc,p-p}}{V_{loaded,p-p}}-1}$ kΩ.

Fig. 2.28   Simplified model
for circuit shown in Fig. 2.27

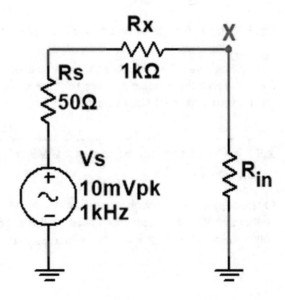

Now remove the capacitor C3 and measure the input resistance again (Fig. 2.29). Compare the obtained value with the one obtained for Fig. 2.25. What is the effect of R4 on the input impedance?

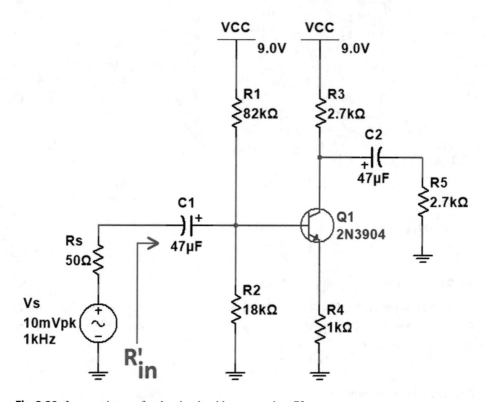

Fig. 2.29 Input resistance for the circuit without capacitor C3

Use hand analysis to calculate the input resistance of circuits shown in Figs. 2.25 and 2.29. Compare the calculated values with the measured ones. Try to find the source of any discrepancy between these values.

2.9   Measurement of Output Resistance for Common Emitter Amplifier

In this experiment we want to measure the output resistance (Fig. 2.30) of a common emitter amplifier. The technique studied in this experiment can be used to measure the output resistance of other type of amplifiers as well.

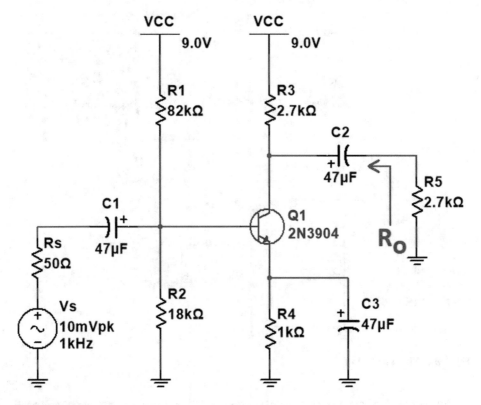

Fig. 2.30 Output resistance

Simple equivalent circuit for Fig. 2.30 is shown in Fig. 2.31. Let's measure the peak-to-peak of voltage source V_o in Fig. 2.31. In order to do this, disconnect the load and measure the peak-to-peak of the waveform on the oscilloscope screen (Fig. 2.32). Measured value is the peak-to-peak of V_o. Lets call the measured value $V_{o,p-p}$.

Fig. 2.31 Simplified model for circuit shown in Fig. 2.30

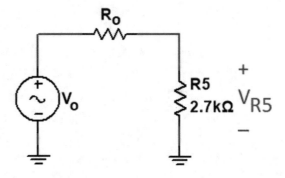

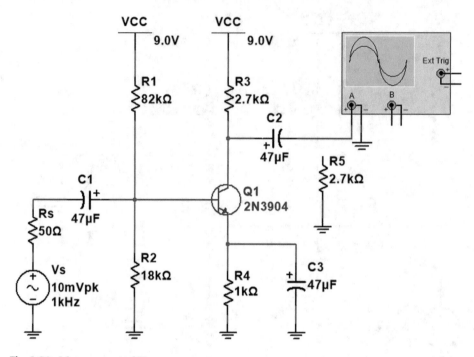

Fig. 2.32 Measurement of V_o

Connect the load and measure the peak-to-peak of the waveform you see on the oscil-loscope screen (Fig. 2.33). Measured value is the peak-to-peak of resistor R5 voltage (Fig. 2.31). Lets call the measured value $V_{R5,p-p}$.

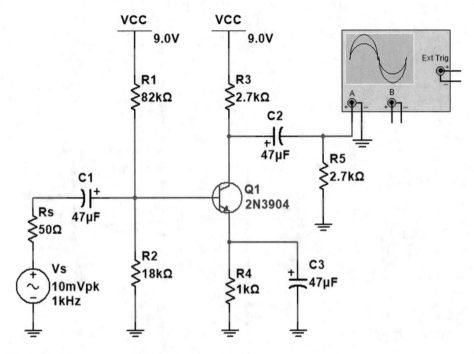

Fig. 2.33 Measurement of VR5

The output resistance (in kΩ) can be calculated as $\frac{R_5}{R_o+R_5}V_{o,p-p} = V_{R5,p-p} \Rightarrow R_o = \left(\frac{V_{o,p-p}}{V_{R5,p-p}} - 1\right) \times R_5 = 2.7\left(\frac{V_{o,p-p}}{V_{R5,p-p}} - 1\right)$ kΩ according to Fig. 2.31. Use this formula to find the output resistance.

Now remove the capacitor C3 (Fig. 2.34) and measure the output impedance. What is the effect of R4 on output impedance?

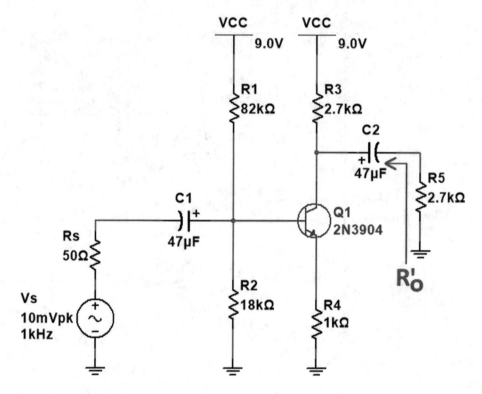

Fig. 2.34 Output resistance

Use hand analysis to calculate the output resistance of circuits shown in Figs. 2.30 and 2.34. Compare the calculated values with the measured ones. Try to find the source of any discrepancy between these values.

2.10   Frequency Response of Common Emitter Amplifier

In this experiment we want to obtain the frequency response of a common emitter amplifier. A complete frequency response is composed of two graphs: Magnitude response and phase response. Only the magnitude response is studied here. The technique studied in this experiment can be used to extract the frequency response of other type of amplifiers as well.

Typical frequency response of an amplifier is shown in Fig. 2.35. It is composed of three regions, low frequency region, mid-band region and high frequency region. The gain in the mid-band region is almost constant. The -3 dB cut-off frequencies are defined as frequencies where mid-band gain decreases to $0.707 \times G_{mid}$. These frequencies are shown with F_L and F_H in Fig. 2.35.

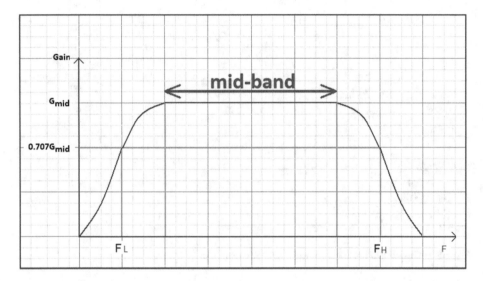

Fig. 2.35 Frequency response of a typical amplifier

Let's get started. Make the circuit shown in Fig. 2.36. Rs shows the output resistance of the function generator Vs. Generate the frequencies shown in Table 2.5 and fill it. Use the Table 2.5 to find the mid-band gain (G_{mid}) of the amplifier. Draw the graph of obtained data with the aid of MATLAB (see Sect. A.7 in Appendix A).

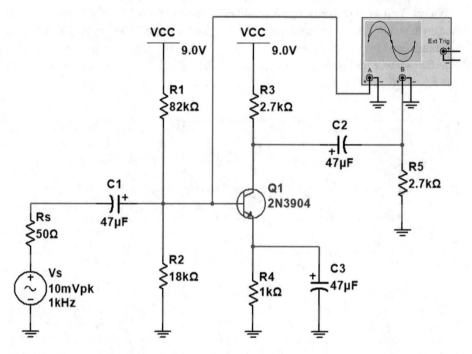

Fig. 2.36 Circuit used to measure the frequency response

Table 2.5 Gain of amplifier shown in Fig. 2.36 for different frequencies

Frequency	100 Hz	200 Hz	500 Hz	1 kHz	10 kHz	50 kHz	100 kHz	200 kHz	500 kHz
Peak-to-peak of output voltage ($V_{pp,out}$)									
Peak-to-peak of input voltage ($V_{pp,in}$)									
$Gain = \dfrac{V_{pp,out}}{V_{pp,in}}$									

Now add a 1 nF capacitor between base and collector (Fig. 2.37) and fill the Table 2.6.

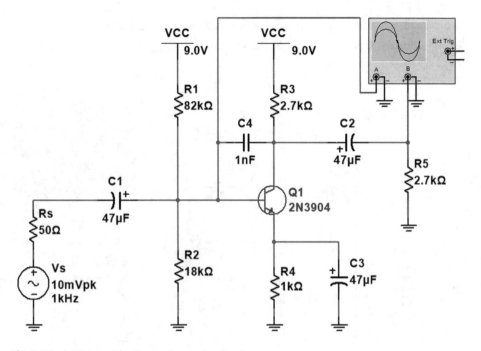

Fig. 2.37 Addition of 1 nF capacitor to the circuit

Table 2.6 Gain of amplifier shown in Fig. 2.37 for different frequencies

Frequency	100 Hz	200 Hz	500 Hz	1 kHz	10 kHz	50 kHz	100 kHz	200 kHz	500 kHz
Peak-to-peak of output voltage ($V_{pp,out}$)									
Peak-to-peak of input voltage ($V_{pp,in}$)									
$Gain = \frac{V_{pp,out}}{V_{pp,in}}$									

Use MATLAB to draw the graph of Tables 2.5 and 2.6 on the same axis. This permits you to compare them with each other easily. What is the effect of addition of C4 (**Hint**: remember the Miller theorem)?

2.11 Common Base Amplifier

This experiment studies the common base amplifier. Make the circuit shown in Fig. 2.38. Rs shows the output resistance of the signal generator Vs. Use a DC voltmeter to measure the DC voltage of base, emitter and collector. Use a DC micro ammeter to measure the base current and calculate the current gain of the transistor using the $\beta = \frac{\frac{V_{CC}-V_C}{R_3}}{I_B} = \frac{V_{CC}-V_C}{R_3 \times I_B}$ formula. Operating point of the transistor is $V_{CEQ} = V_C - V_E$ and $I_{CQ} = \frac{V_{CC}-V_C}{R_3}$.

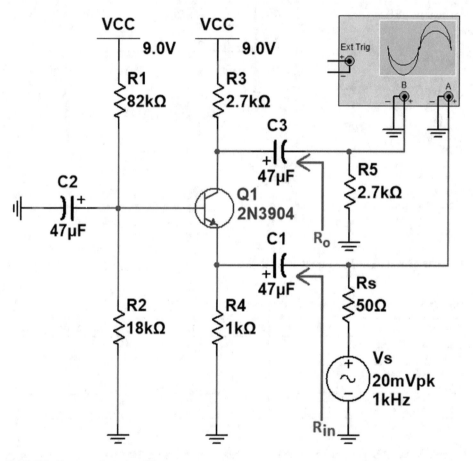

Fig. 2.38 Common base amplifier

Measure the voltage gain, input resistance and output resistance of the amplifier and compare it with the values that are obtained using hand analysis. Try to find the reason of any discrepancy.

2.12   Common Collector Amplifier

This experiment studies the common collector amplifier. Make the circuit shown in Fig. 2.39. Rs shows the output resistance of the signal generator Vs. Use a DC voltmeter to measure the DC voltage of base, emitter and collector. Use a DC micro ammeter to measure the base current and calculate the current gain of the transistor using the $\beta \approx \frac{\frac{V_E}{R_3}}{I_B} = \frac{V_E}{R_3 \times I_B}$ formula. Operating point of the transistor is $V_{CEQ} = V_{CC} - V_E$ and $I_{CQ} \approx \frac{V_E}{R_3}$.

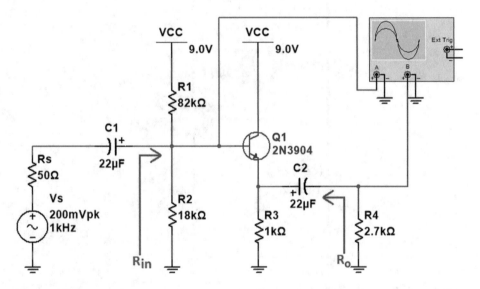

Fig. 2.39 Common collector amplifier

Measure the voltage gain, input resistance (use Rx = 10 kΩ, see Sect. 2.8) and output resistance of the amplifier and compare it with the values that are obtained using hand analysis. Try to find the reason of any discrepancy.

2.13   Darlington Configuration

Input impedance of a common collector amplifier shown in Fig. 2.39 is a function of transistor's current gain (β). Increase in β means bigger input impedance which is a desired thing. The β can be increased with the aid Darlington configuration. This experiment studies the properties of Darlington configuration.

Make the circuit shown in Fig. 2.40.

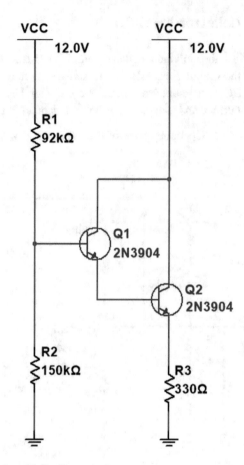

Fig. 2.40 Voltage divider with Darlington pair

Use a micro ammeter to measure the base current I_B (Fig. 2.41).

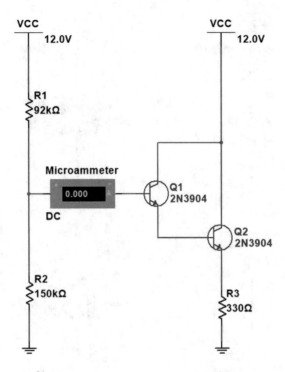

Fig. 2.41 Measurement of base current

Use a milli ammeter to measure the collector current I_C (Fig. 2.42). Measure the equivalent current gain with the aid of $\beta = \frac{I_C}{I_B}$. Figure 2.43 shows the equivalent of Fig. 2.40. Hypothetical transistor Q has gain of $\beta = \frac{I_C}{I_B}$ and base-emitter voltage drop of around 2 times a normal transistor.

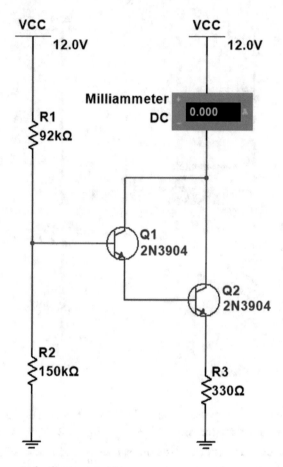

Fig. 2.42 Measurement of collector current

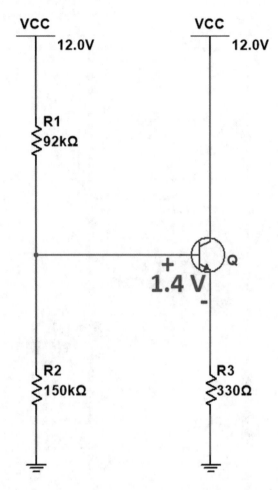

Fig. 2.43 Voltage drop between base and emitter is 1.4 V for Darlington pair

Make the circuit shown in Fig. 2.44. Use a DC voltmeter to measure the voltage of different nodes. Use a micro ammeter to measure the I_B and I_C. Calculate the current gain using the $\frac{I_C}{I_B}$ formula.

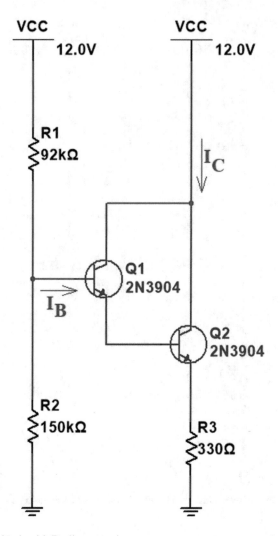

Fig. 2.44 Sample circuit with Darlington pair

Change the circuit to what shown in Fig. 2.45. Rs1 is the output resistance of the signal generator V1. Rs2 is a resistor which helps us to measure the input resistance of the amplifier. Measure the voltage gain (i.e., peak-to-peak of Channel B divided by peak-to-peak of Channel A) of the amplifier shown in Fig. 2.45.

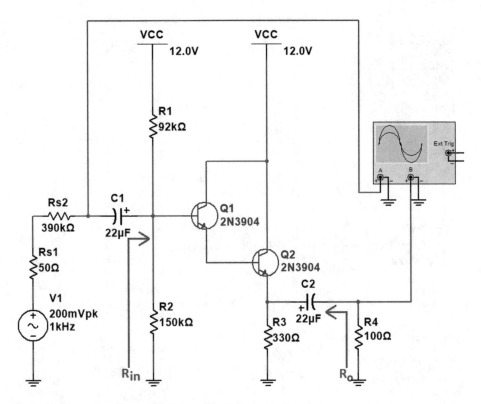

Fig. 2.45 Common collector amplifier with Darlington pair

Let's measure the input resistance of the amplifier as well. Disconnect the connection between signal generator V1 and Rs2 (Fig. 2.46) and measure the pea-to-peak of voltage source V1. Call the measured value $V_{1,pp}$.

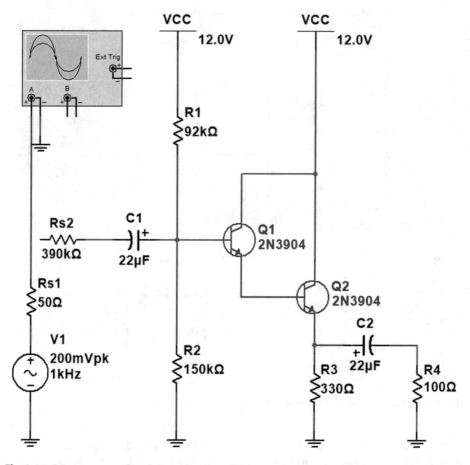

Fig. 2.46 Measurement of peak-to-peak value of V1

Now make the connection between voltage source V1 and Rs2 (Fig. 2.47). Measure the peak-to-peak of voltage across the Rs2. Call the measured value $V_{Rs2,pp}$.

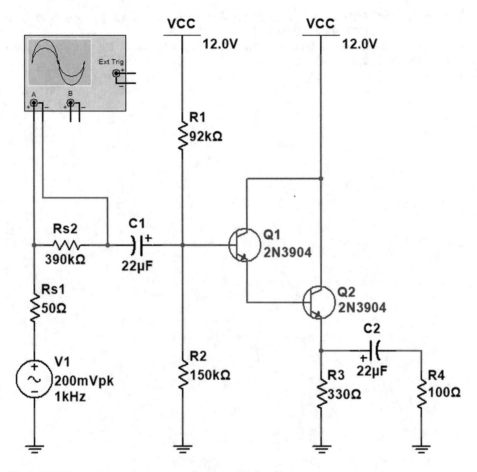

Fig. 2.47 Measurement of peak-to-peak value of Rs2 voltage

Value of R_{in} can be calculated with the aid of $\frac{R_{in}}{R_{in}+390k+50} \times V_{1,pp} = V_{Rs2,pp} \Rightarrow$ $R_{in} = \frac{390.05}{\frac{V_{1,pp}}{V_{2,pp}}-1}$ kΩ formula. Compare the obtained value with the input resistance of common collector amplifier shown in Fig. 2.39.

References for Further Study

1. Asadi F., Essential Circuit Analysis using Proteus, Springer, 2022. DOI: https://doi.org/10.1007/978-981-19-4353-9
2. Asadi F., Essential Circuit Analysis using LTspice, Springer, 2022. DOI: https://doi.org/10.1007/978-3-031-09853-6

3. Asadi F., Essential Circuit Analysis using NI Multisim and MATLAB, Springer, 2022. DOI: https://doi.org/10.1007/978-3-030-89850-2
4. Asadi F., Electric Circuit Analysis with EasyEDA, Springer, 2022. DOI: https://doi.org/10.1007/978-3-031-00292-2

MOSFET Transistor Amplifiers

3

3.1 Introduction

Previous chapter studied the BJT transistor. This chapter studies Metal Oxide Semiconductor Field Effect Transistor (MOSFET) which is another important type of transistor. Here is a brief comparison between BJT and MOSFET: In BJT, charge carriers are both electrons and holes while in MOSFET, either electrons or holes act as charge carriers depending on the type of channel between source and drain. Switching speed of MOSFET is higher than BJT. BJT is a current controlled device while MOSFET is a voltage controlled device. MOSFET's are used more than BJT's in most of the applications. BJT doesn't suffer from the electrostatic discharge while MOSFET suffer from electrostatic discharge.

3.2 Measurement of Threshold Voltage

In this experiment we will measure the threshold voltage of an NMOS transistor. The transistor used in this experiment is 2N7000. Pinout of 2N7000 is shown in Fig. 3.1.

© The Author(s), under exclusive license to Springer Nature Switzerland AG 2023   79
F. Asadi, *Analog Electronic Circuits Laboratory Manual*, Synthesis Lectures on Electrical Engineering, https://doi.org/10.1007/978-3-031-25122-1_3

Fig. 3.1 Pinout of 2N7000
NMOS transistor

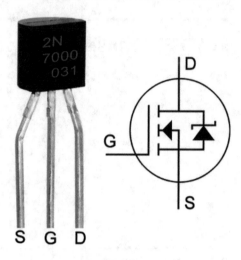

Let's get started. Make the circuit shown in Fig. 3.2. In the beginning of the experiment set the voltage of gate (VG) to minimum, i.e., around 0 V. Then start to increase the gate voltage slowly (voltage of gate is controlled with potentiometer P1). Stop increasing the voltage when the milli ammeter showed a current in the range of 0.5–1 mA. Measure the voltage of the gate. Measured voltage is the threshold voltage of the transistor. For 2N7000, the threshold voltage has minimum of 1 V and maximum of 3 V. Typical value is 2 V.

Fig. 3.2 Measurement of
threshold voltage

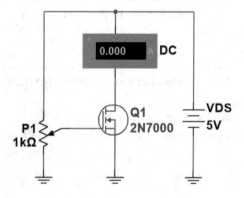

3.3   Current Voltage Characteristic of MOS Transistor

In this experiment we want to obtain the current voltage characteristic of 2N7000 transistor. Measure the threshold of your transistor with the aid of method explained in the previous experiment. After measuring the threshold, make the circuit shown in Fig. 3.3.

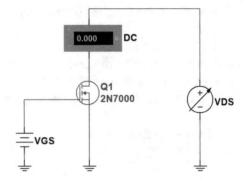

Fig. 3.3 Circuit to extract the I–V characteristic of 2N7000 NMOS transistor

Set the VGS equal to 0.5, 2.5 and 4 V and fill the Tables 3.1, 3.2 and 3.3.

Table 3.1 Drain current for VGS = 0.5 V

VDS	0 V	0.2 V	0.4 V	0.6 V	0.8 V	1 V	1.5 V	2 V	3 V	4 V	6 V
Drain current											

Table 3.2 Drain current for VGS = 2.5 V

VDS	0 V	0.2 V	0.4 V	0.6 V	0.8 V	1 V	1.5 V	2 V	3 V	4 V	6 V
Drain current											

Table 3.3 Drain current for VGS = 4 V

VDS	0 V	0.2 V	0.4 V	0.6 V	0.8 V	1 V	1.5 V	2 V	3 V	4 V	6 V
Drain current											

Use MATLAB to draw the graph of measured data and determine the triode and saturation regions. In the saturation region, the current of the transistor follows the $I_D = K_n(V_{gs} - V_{th})^2$ relationship (effect of channel length modulation is neglected). I_D, V_{gs} and V_{th} show the drain current, gate-source voltage and threshold voltage, respectively. Use Table 3.3 to estimate value of K_n.

3.4   Common Source Amplifier

Common source amplifier is studied in this experiment. For amplification purpose, the transistor must work in the saturation region. Remember that in the saturation region, the current of the transistor follows the $I_D = K_n(V_{gs} - V_{th})^2$ relationship (effect of channel length modulation is neglected). I_D, V_{gs} and V_{th} show the drain current, gate-source voltage and threshold voltage, respectively. Value of transistor transconductance is given by $\frac{\partial I_D}{\partial V_{gs}} = \frac{2I_D}{(V_{gs} - V_{th})}$.

Let's get started. Make the circuit shown in Fig. 3.4. Use the potentiometer to set the drain current equal to 20 mA. When drain current is 20 mA, VRS = 2 V. So you can monitor voltage of resistor RS in order to see whether drain current is 20 mA. Use a DC voltmeter to measure the voltage of all of the nodes of the circuit.

Fig. 3.4 Biasing circuit for 2N7000 transistor

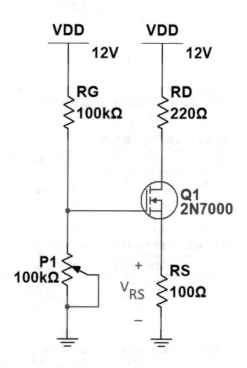

After obtaining the drain current of 20 mA, change the circuit to what shown in Fig. 3.5. RO shows the output resistance of the signal generator. V1 and RO + R1 represents the Thevenin equivalent of any circuit that drives the amplifier. Measure the amplifier gain (i.e., peak-to-peak of Channel B divided by peak-to-peak of Channel A). Use the technique shown in Sect. 2.8 to measure the input impedance of the amplifier (Use R1 as RX). Measure the output resistance of the amplifier as well.

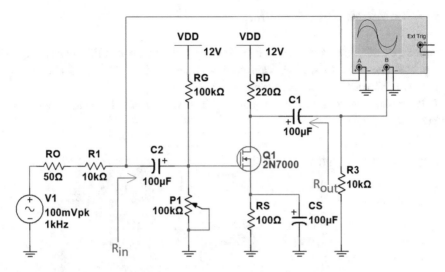

Fig. 3.5 Common source amplifier

Remove the capacitor CS (Fig. 3.6). Measure the gain, input resistance and output resistance for Fig. 3.6.

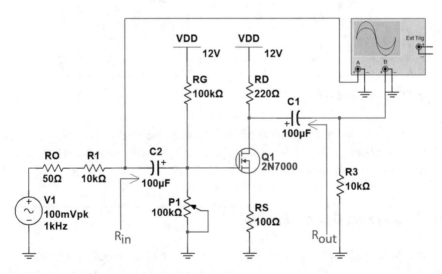

Fig. 3.6 Capacitor CS is removed from the circuit

Do a hand analysis for the given circuits and calculate the voltage gain, input resistance and output resistance. Compare the result of hand analysis with the measured values. Try to explain reason of any discrepancy.

3.5   Common Gate Amplifier

Common gate amplifier is studied in this experiment. Make the circuit shown in Fig. 3.7. Use potentiometer P1 to set the drain current equal to 20 mA (when drain current is 20 mA, the DC voltage of source terminal is 2 V).

After setting the drain current equal to 20 mA, measure the voltage gain, input resistance and output resistance.

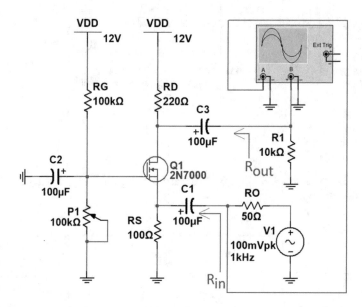

Fig. 3.7 Common gate amplifier

Do a hand analysis for the given circuit and calculate the voltage gain, input resistance and output resistance. Compare the result of hand analysis with the measured values. Try to explain reason of any discrepancy.

3.6   Common Drain Amplifier

Common drain amplifier is studied in this experiment. Make the circuit shown in Fig. 3.8. Use potentiometer P1 to set the drain current equal to 20 mA (when drain current is 20 mA, the DC voltage of source terminal is 2 V).

After setting the drain current equal to 20 mA, measure the voltage gain, input resistance and output resistance.

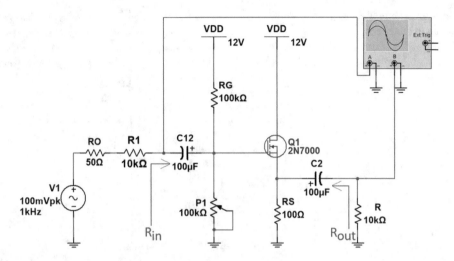

Fig. 3.8 Common drain amplifier

Do a hand analysis for the given circuit and calculate the voltage gain, input resistance and output resistance. Compare the result of hand analysis with the measured values. Try to explain reason of any discrepancy.

3.7   MOSFET as a Switch

In this experiment we will use a MOSFET as a switch. Let's get started. Make the circuit shown in Fig. 3.9. RO shows the output resistance of the signal generator V1. Using a 150 Ω 1W resistor is recommended in this circuit. Waveform generated by signal generator V1 is shown in Fig. 3.10. According to the datasheet, the IRFZ has a threshold voltage which is between 2 and 4 V. When gate source voltage is bigger than the threshold voltage, the MOSFET act as a closed switch. Therefore, applying 5 V to gate-source of IRFZ 44 cause the device to be closed (since 5 V is bigger than the maximum threshold voltage which in this case is 4 V). Applying 0 V cause the device to be opened as well (since 0 V is less than the minimum threshold voltage which in this case is 2 V).

Fig. 3.9 MOSFET act as a
switch in this circuit

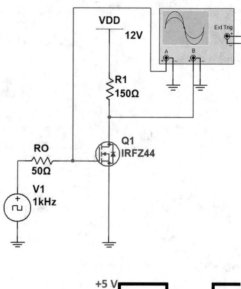

Fig. 3.10 Waveform of input
voltage source V1

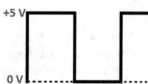

Observe the waveforms of Channel A and B simultaneously and draw them in your lab notebook. Measure the drain-source voltage when the MOSFET is on and calculate the average conduction loss of the MOSFET. Measure the RMS and average value of the resistor R1 voltage. Do a hand analysis and compare your results with the measured values.

References for Further Study

1. Asadi F., Essential Circuit Analysis using Proteus, Springer, 2022. DOI: https://doi.org/10.1007/978-981-19-4353-9
2. Asadi F., Essential Circuit Analysis using LTspice, Springer, 2022. DOI: https://doi.org/10.1007/978-3-031-09853-6
3. Asadi F., Essential Circuit Analysis using NI Multisim and MATLAB, Springer, 2022. DOI: https://doi.org/10.1007/978-3-030-89850-2
4. Asadi F., Electric Circuit Analysis with EasyEDA, Springer, 2022. DOI: https://doi.org/10.1007/978-3-031-00292-2

Current Sources and Differential Pair

4

4.1 Introduction

Current source and differential pair are two important circuits with many applications. For instance, transistors inside of an Integrated Circuit (IC) are biased with the aid of current sources and input stage of an Op amp is a differential amplifier. This chapter studies important current sources and BJT differential pairs.

4.2 Current Mirror

In this experiment we will study the current mirror circuit. Consider the current mirror circuit shown in Fig. 4.1 (both transistors are assumed to be the same). It can be shown that $I_o = \frac{1}{1+\frac{2}{\beta}} I_{REF} \approx I_{REF}$.

Fig. 4.1 Current mirror circuit

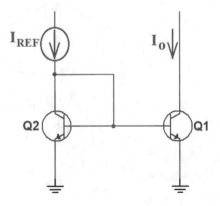

© The Author(s), under exclusive license to Springer Nature Switzerland AG 2023
F. Asadi, *Analog Electronic Circuits Laboratory Manual*, Synthesis Lectures on Electrical Engineering, https://doi.org/10.1007/978-3-031-25122-1_4

Make the circuit shown in Fig. 4.2. Use a DC voltmeter to measure the voltage of nodes. Calculate the I_{REF} and I_o using the $I_{REF} = \frac{V_{CC}-V_{C2}}{R_1}$ and $I_o = \frac{V_{CC}-V_{C1}}{R_{load}}$ formulas. V_{C1} and V_{C2} show collector voltage of transistor Q1 and Q2, respectively. I_{REF} is around 1.95 mA since $I_{REF} = \frac{V_{CC}-V_{C2}}{R_1} \approx \frac{5-0.7}{2.2k} = 1.95\,\mathrm{mA}$.

Fig. 4.2 Current mirror circuit used in the experiment

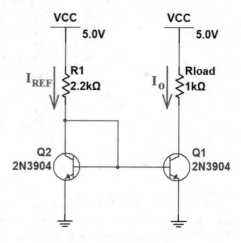

4.3   Widlar Current Source

In this experiment we will study the Widlar current source (Fig. 4.3). Both transistors are assumed to be the same. It can be shown that $I_o R_E = V_T \ln(\frac{I_{REF}}{I_o})$. V_T shows the thermal voltage and for room temperature, $V_T = 25.8\,\mathrm{mV}$.

Fig. 4.3   Widlar current source

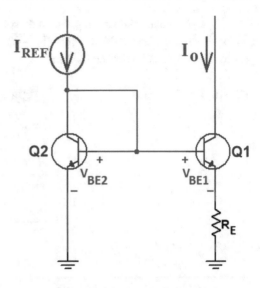

Make the circuit shown in Fig. 4.4. Use a DC voltmeter to measure the voltage of nodes. Calculate the I_{REF} and I_o using the $I_{REF} = \frac{V_{CC} - V_{C2}}{R_1}$ and $I_o = \frac{V_{CC} - V_{C1}}{R_{load}}$ formulas. V_{C1} and V_{C2} show collector voltage of transistor Q1 and Q2, respectively. I_{REF} is around 1.95 mA since $I_{REF} = \frac{V_{CC} - V_{C2}}{R_1} \approx \frac{5 - 0.7}{2.2k\Omega} = 1.95$ mA.

Fig. 4.4   Widlar current source
used in the experiment

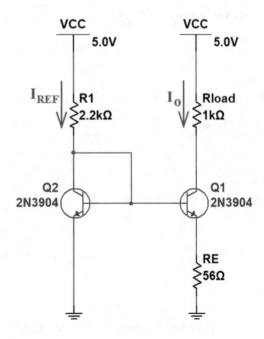

MATLAB code shown in Fig. 4.5 solves the nonlinear equation $I_o R_E = V_T \ln(\frac{I_{REF}}{I_o}) \Rightarrow 56 I_o = 0.0258 \times \ln(\frac{1.95 \times 10^{-3}}{I_o})$. According to Fig. 4.5, the output current I_O in Fig. 4.4 must be around 0.57 mA.

Fig. 4.5 MATLAB code

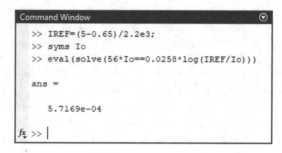

```
Command Window
>> IREF=(5-0.65)/2.2e3;
>> syms Io
>> eval(solve(56*Io==0.0258*log(IREF/Io)))

ans =

    5.7169e-04

fx >> |
```

4.4   Wilson Current Mirror

In this experiment we will study the Wilson current source (Fig. 4.6). All of the transistors are assumed to be the same. It can be shown that $I_o \approx \frac{1}{1+\frac{2}{\beta^2}} I_{REF}$.

Fig. 4.6 Wilson current mirror

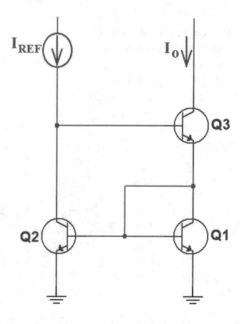

Make the circuit shown in Fig. 4.7. Use a DC voltmeter to measure the voltage of nodes. Calculate the I_{REF} and I_o using the $I_{REF} = \frac{V_{CC} - V_{B3}}{R_1}$ and $I_o = \frac{V_{CC} - V_{C3}}{R_{load}}$ formulas.

V_{B3} and V_{C3} show base and collector voltages of transistor Q3, respectively. I_{REF} is around 1.64 mA since $I_{REF} = \frac{V_{CC}-V_{B3}}{R_1} \approx \frac{5-0.7-0.7}{2.2k} = 1.64\text{mA}$.

Fig. 4.7 Wilson current mirror used in the experiment

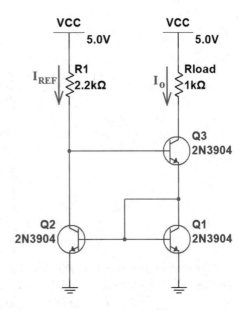

4.5   Current Source with Op Amp

In this experiment we will study the Op amp based current sources. Consider the circuit shown in Fig. 4.8. Potential of node P is controlled with the aid of potentiometer P1. Presence of negative feedback makes the potential of node N equal to node P. Therefore, the current that pass through R3 equals to $\frac{V_N}{R_3} = \frac{V_P}{R_3}$. The + and − terminals of an Op amp draws almost zero current. So, current through load resistor Rload equals to the current through resistor R3, i.e., $I_o = \frac{V_P}{R_3}$.

Fig. 4.8 Current source with
Op Amp

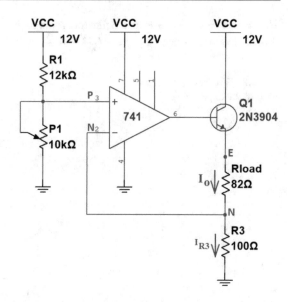

Make the circuit shown in Fig. 4.8. Turn the potentiometer until potential of node P reaches 2 V. Measure the potential of node N and E. Calculate the output current using the $I_o = \frac{V_E - V_N}{Rload}$ formula. Compare the calculated value with $I_{R3} = \frac{V_N}{R3}$. Repeat the experiment for the case that potential of node P equals to 5 V.

4.6   Differential Pair (Resistive Bias)

In this experiement we will study the common mode and differential mode gains of the differential pair shown in Fig. 4.9. Transistors of this circuit are biased with the aid of resistor RE. Current of collector of Q1 and Q2 is around $\frac{-0.7-(-12)}{\frac{10}{2}} = 0.57 \text{mA}$.

Fig. 4.9 Differential pair

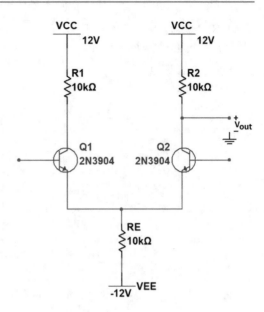

Ideally, a differential amplifier takes the voltages, V_+ and V_- on its two inputs and produces an output voltage $V_o = A_d(V_+ - V_-)$, where A_d is the differential gain. However, the output of a real differential amplifier is better described as: $V_o = A_d(V_+ - V_-) + \frac{1}{2}A_{cm}(V_+ + V_-)$.

Make the circuit shown in Fig. 4.10. Use a DC voltmeter to measure the voltage of nodes. Calculate the collector currents using $I_{C1} = \frac{V_{CC} - V_{C1}}{R_1}$ and $I_{C2} = \frac{V_{CC} - V_{C2}}{R_2}$ as well. V_{C1} and V_{C2} shows the voltage of Q1 and Q2's collector, respectively. Compare the obtained results with the results of hand analysis.

Fig. 4.10 Measurement of quiescent DC operating points

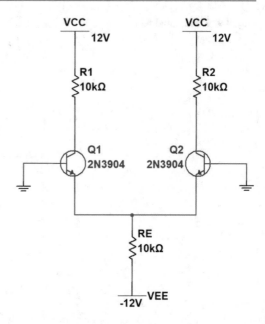

Change the circuit to what shown in Fig. 4.11. Rs shows the output resistance of the signal generator V1. Coupling of both channels are set to AC. Measure the gain (peak-to-peak of channel B divided by peak-to-peak of Channel A). Obtained value is the common mode gain A_{CM}.

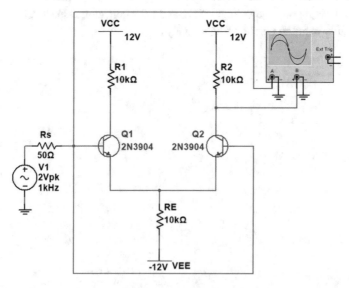

Fig. 4.11 Measurement of common mode gain

Change the circuit to what shown in Fig. 4.12. R3 and R4 make a voltage divider, therefore a small voltage (around 20 mV) reaches to base of Q1 and the output doesn't saturate. Coupling of both channels are set to AC. Measure the gain (peak-to-peak of channel B divided by peak-to-peak of Channel A) and show it with A_{DO}. The differential mode gain $A_{DM} = A_{DO} - \frac{A_{CM}}{2}$.

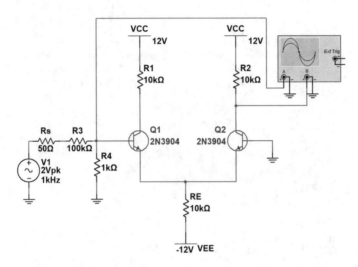

Fig. 4.12 Measurement of differential gain

The Common Mode Rejection Ratio (CMRR) can be calculated using $20\log(\frac{A_{DM}}{A_{CM}})$.

4.7 Differential Pair (Current Source Bias)

In this experiement we wil study the common mode and differential mode gains of the differential pair shown in Fig. 4.13. Transistors of this circuit are biased with the aid of a current mirror composed of Q3 and Q4. Current of collector of Q3 is around $\frac{12-(-11.3)}{22} = 1.06$ mA. Therefore, current of collector Q1 and Q2 is around half of this value (i.e., 0.53 mA).

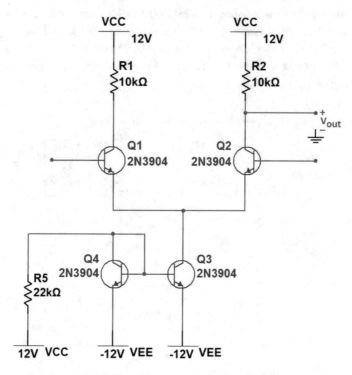

Fig. 4.13 Current source biased differential pair

Make the circuit shown in Fig. 4.14. Use a DC voltmeter to measure the voltage of nodes. Calculate the collector currents using $I_{C1} = \frac{V_{CC}-V_{C1}}{R_1}$ and $I_{C2} = \frac{V_{CC}-V_{C2}}{R_2}$ as well. V_{C1} and V_{C2} shows the voltage of Q1 and Q2's collector, respectively. Compare the obtained results with the results of hand analysis.

Fig. 4.14 Measurement of
quiescent DC operating points

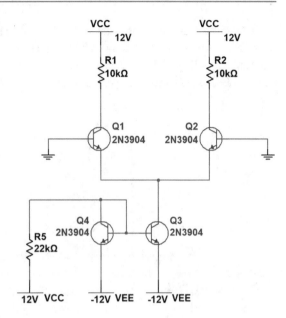

Change the circuit to what shown in Fig. 4.15. Rs shows the output resistance of signal generator V1. Coupling of both channels are set to AC. Measure the gain (peak-to-peak of channel B divided by peak-to-peak of Channel A). Obtained value is the common mode gain A_{CM}.

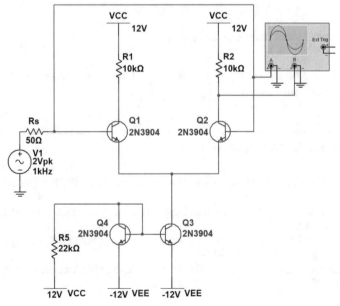

Fig. 4.15 Measurement of common mode gain

Now change the circuit to what shown in Fig. 4.16. R3 and R4 make a voltage divider, therefore a small voltage (around 20 mV) reaches to base of Q1 and the output doesn't saturate. Coupling of both channels are set to AC. Measure the gain (peak-to-peak of channel B divided by peak-to-peak of Channel A). Obtained value is the differential mode gain A_{DM}.

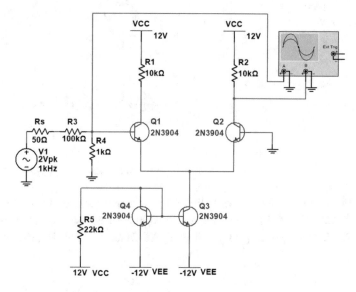

Fig. 4.16 Measurement of differential mode gain

The CMRR can be calculated using $20\log(\frac{A_{DM}}{A_{CM}})$. Compare the obtained number with the CMRR of previous experiment.

References for Further Study

1. Asadi F., Essential Circuit Analysis using Proteus, Springer, 2022. DOI: https://doi.org/10.1007/978-981-19-4353-9
2. Asadi F., Essential Circuit Analysis using LTspice, Springer, 2022. DOI: https://doi.org/10.1007/978-3-031-09853-6
3. Asadi F., Essential Circuit Analysis using NI Multisim and MATLAB, Springer, 2022. DOI: https://doi.org/10.1007/978-3-030-89850-2
4. Asadi F., Electric Circuit Analysis with EasyEDA, Springer, 2022. DOI: https://doi.org/10.1007/978-3-031-00292-2

Transistor Feedback Amplifiers

5

5.1 Introduction

Feedback is an important concept in electronics. Suitable type of negative feedback permits an amplifier to obtain a higher bandwidth, lower output impedance, higher input impedance and reduce the nonlinear distortions. Gain of amplifiers which use negative feedback is less sensitive to variations in the values of circuit components and temperature.

Under certain conditions, the negative feedback can become positive and of such a magnitude as to cause oscillations. Danger of such an oscillation decreases considerably if the feedback loop is designed carefully.

This chapter studies the negative feedback in transistor amplifiers.

5.2 Voltage Series Feedback

Make the circuit shown in Fig. 5.1. This amplifier uses the voltage series feedback. Set the coupling of both channels to AC and measure the voltage gain (i.e., peak-to-peak of output divided by peak-to-peak of input) of the amplifier.

© The Author(s), under exclusive license to Springer Nature Switzerland AG 2023   99
F. Asadi, *Analog Electronic Circuits Laboratory Manual*, Synthesis Lectures on Electrical
Engineering, https://doi.org/10.1007/978-3-031-25122-1_5

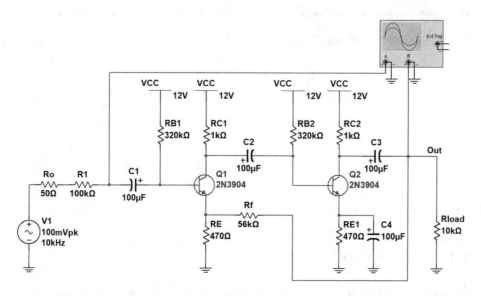

Fig. 5.1 An amplifier with voltage series feedback

Use the technique studied in Sect. 2.8 to measure the input resistance R_{in} (Fig. 5.2). R1 in Fig. 5.2 plays the role of Rx in Sect. 2.8.

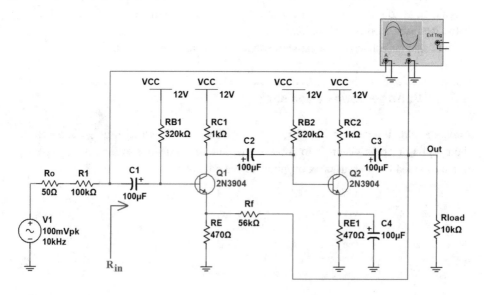

Fig. 5.2 Input resistance

Let's measure the output resistance (Fig. 5.3) as well. Measure the output peak-to-peak value for Fig. 5.3. Call the measured value $V_{pp,10k}$.

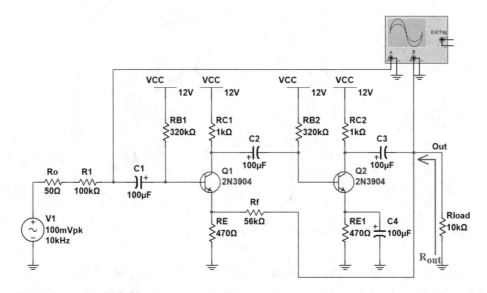

Fig. 5.3 Output resistance

The simple model for Fig. 5.3 is shown in Fig. 5.4. This model has two unknowns: V_{out} and R_{out}. A system of two equations can help us to determine these values. Let's get started. According to Fig. 5.4, $\frac{10}{10+R_{out}} \times V_{out} = V_{pp,10k}$.

Fig. 5.4 Simplified model of Fig. 5.3

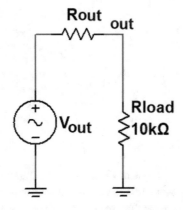

Now add another 10 kΩ resistor to the output (Fig. 5.5). This decreases the output load to 5 kΩ. Measure the output peak-to-peak value for Fig. 5.5. Call the measured value $V_{pp,5k}$.

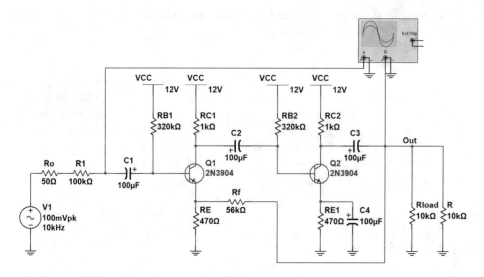

Fig. 5.5 Addition of a 10 kΩ resistor to output

The simple model for Fig. 5.5 is shown in Fig. 5.6. According to Fig. 5.6, $\frac{5}{5+R_{out}} \times$
$V_{out} = V_{pp,5k}$. Let's define $\alpha = \frac{V_{pp,10k}}{V_{pp,5k}}$. The output resistance Rout can be found using
the $R_{out} = \frac{\alpha-1}{2-\alpha} \times 10k$" formula.

Fig. 5.6 Simplified model for
Fig. 5.5

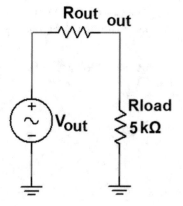

Do a hand analysis for the circuit shown in Fig. 5.1 and determine the voltage gain,
input resistance and output resistance. Compare the obtained values with the measured
values. Try to find the reason of any discrepancy.

5.3   Voltage Shunt Feedback

Make the circuit shown in Fig. 5.7. This amplifier uses the voltage shunt feedback. Coupling of both channels are set to AC. Measure the voltage gain, input resistance and output resistance of this amplifier using the procedure explained in the previous experiment.

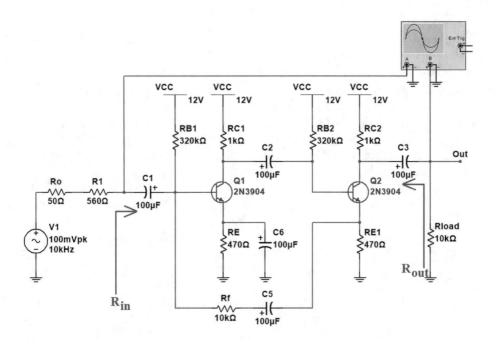

Fig. 5.7 An amplifier with voltage shunt feedback

Do a hand analysis for the circuit shown in Fig. 5.7 and determine the voltage gain, input resistance and output resistance. Compare the obtained values with the measured values. Try to find the reason of any discrepancy.

References for Further Study

1. Asadi F., Essential Circuit Analysis using Proteus, Springer, 2022. DOI: https://doi.org/10.1007/978-981-19-4353-9
2. Asadi F., Essential Circuit Analysis using LTspice, Springer, 2022. DOI: https://doi.org/10.1007/978-3-031-09853-6
3. Asadi F., Essential Circuit Analysis using NI Multisim and MATLAB, Springer, 2022. DOI: https://doi.org/10.1007/978-3-030-89850-2
4. Asadi F., Electric Circuit Analysis with EasyEDA, Springer, 2022. DOI: https://doi.org/10.1007/978-3-031-00292-2

Op Amp Circuits and 555 Timer IC

6

6.1 Introduction

An Operational Amplifier, or Op Amp for short, is fundamentally a voltage amplifying device designed to be used with external feedback components such as resistors and capacitors between its output and input terminals. These feedback components determine the resulting function or "operation" of the amplifier and by virtue of the different feedback configurations whether resistive, capacitive or both, the amplifier can perform a variety of different operations, giving rise to its name of "Operational Amplifier".

All of the circuits in this chapter are designed with the 741 Op Amp. Pins of 741 Op Amp are shown in Fig. 6.1.

Fig. 6.1 Pin out of 741 Op amp

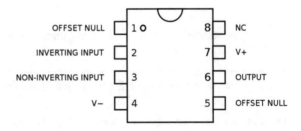

You can connect the supply directly to the IC (Fig. 6.2) and your circuit works. However, it is a good idea to add capacitors to the IC's supply line (Fig. 6.3). Such capacitors must be placed close to the IC and they provide a small source of power storage close to the IC that can buffer against voltage sags. We don't use the capacitors in the circuits of this chapter for the sake of simplicity.

© The Author(s), under exclusive license to Springer Nature Switzerland AG 2023   105
F. Asadi, *Analog Electronic Circuits Laboratory Manual*, Synthesis Lectures on Electrical Engineering, https://doi.org/10.1007/978-3-031-25122-1_6

Fig. 6.2 Supply terminals of
741

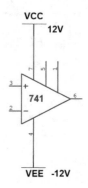

Fig. 6.3 Capacitor C1 and C2
are connected to the supply
lines

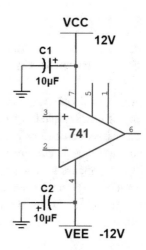

This chapter studies some of important circuits that can be made with Op Amps.

6.2   Measurement of Saturation Voltage

In this experiment we want to measure the maximum and minimum that output can take.
Consider the circuit shown in Fig. 6.4. In this figure positive terminal of the Op Amp is
connected to +6 V and negative terminal is connected to the ground (i.e. 0 V). Such a
big positive voltage difference force the Op Amp to be saturated. Output goes toward the
positive rail voltage.

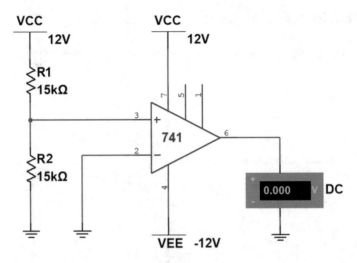

Fig. 6.4 Circuit for measurement of positive saturation voltage

In Fig. 6.5 positive terminal of the Op Amp is connected to −6 V and negative terminal is connected to the ground. Such a big negative voltage difference force the Op Amp to be saturated. Output goes toward the negative rail voltage.

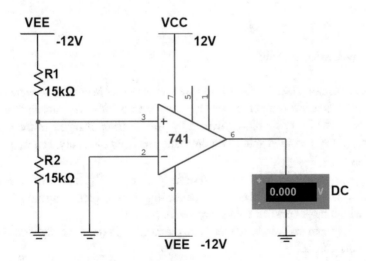

Fig. 6.5 Circuit for measurement of negative saturation voltage

Make the circuit shown in Figs. 6.4 and 6.5. Measure the output voltages and compare them with VCC and VEE.

6.3   Non Inverting Amplifier

A non-inverting amplifier is shown in Fig. 6.6. In this circuit $A_V = \frac{V_{out}}{V_{in}} = 1 + \frac{R_1}{R_2}$.

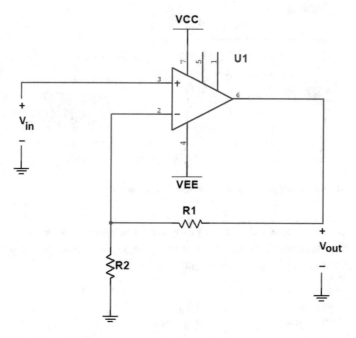

Fig. 6.6 Non-inverting amplifier

Make the circuit shown in Fig. 6.7. RO shows the output resistance of signal generator V1. Measure the phase difference between input and output. Measure the voltage gain (i.e. peak-to-peak of output voltage divided by peak-to-peak of input) of the amplifier as well. Compare the measured value with the value predicted by theory, i.e., $A_v = 1 + \frac{R_1}{R_2} = 1 + \frac{10\ k\Omega}{1\ k\Omega} = 11$.

Observe the input and output simultaneously. Note that presence of negative feedback reduces the nonlinear effects and force the amplifier to behave linearly, i.e., input is amplified with almost constant gain everywhere.

Increase the input amplitude and measure the maximum input amplitude that cause no clipping in the output.

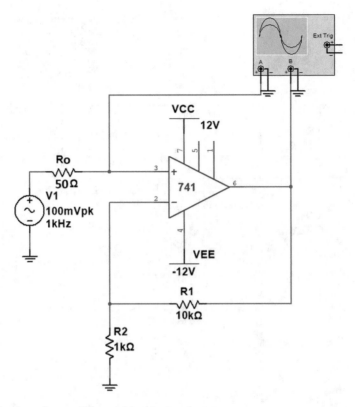

Fig. 6.7 Non-inverting amplifier used in this experiment

The amplifier shown in Fig. 6.6 is a DC coupled amplifier. It can amplify DC signals as well. Let's test the amplifier with DC signals. Make the circuit shown in Fig. 6.8. Turn the potentiometer until voltage of positive terminal of the Op Amp (Channel A) reaches 0.2 V. Measure the output voltage (Channel B). Calculate the voltage gain using output voltage divided by input voltage. Repeat the experiment with input voltage of 0.3 V as well.

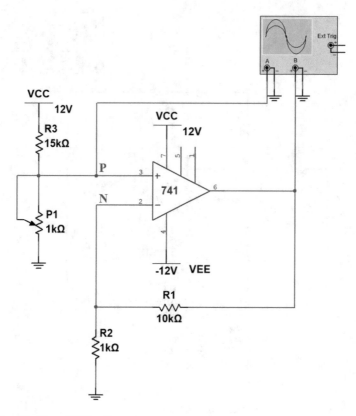

Fig. 6.8 Amplification of DC voltages

6.4   Nulling the Output Offset Voltage

This experiment shows how to neutralize the output offset voltage. Let's get started. Make
the non-inverting amplifier shown in Fig. 6.9.

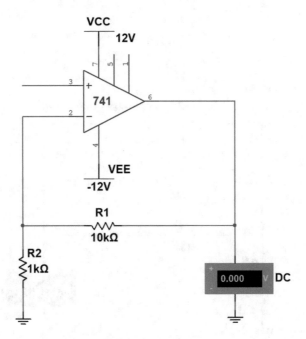

Fig. 6.9 Non-inverting amplifier used in this experiment

Connect the input of the amplifier to the ground (Fig. 6.10) and pay attention to the output voltmeter. From theoretical point of view, input signal to the amplifier in Fig. 6.10 is zero volts therefore output must be zero as well. However, generally few tenth of milli volt are measured in the output despite of zero input. This section shows how to neutralize that small offset voltage.

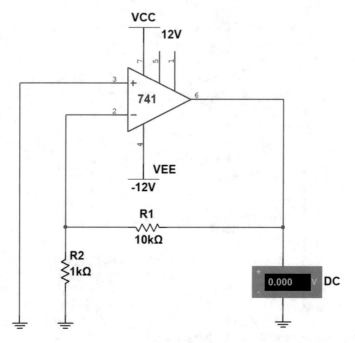

Fig. 6.10   Positive terminal of the Op Amp is connected to ground

   If the milli voltmeter in Fig. 6.10 shows 0 V, then you are very lucky and there is no need to do anything. However, be aware that this case is very rare!

   When the milli voltmeter in Fig. 6.10 shows a few tenth of milli volt, you can add a 10 kΩ potentiometer to the pin 1 and 5 of the Op Amp (Fig. 6.11) and turn the potentiometer until the milli voltmeter shows 0 V. Purpose of pin 1 and 5 of 741 is offset nulling (Fig. 6.1).

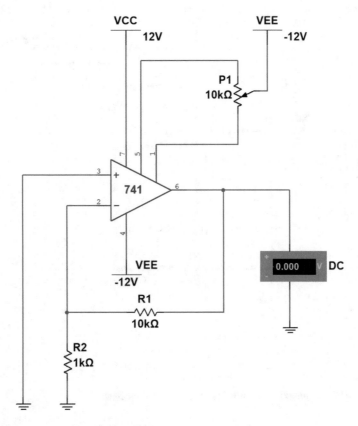

Fig. 6.11 Potentiometer P1 nulls the offset

After setting the potentiometer P1 and obtaining zero output for zero input, you can apply the input signal to the amplifier (Fig. 6.12). RO in Fig. 6.12 shows the output impedance of signal generator V1.

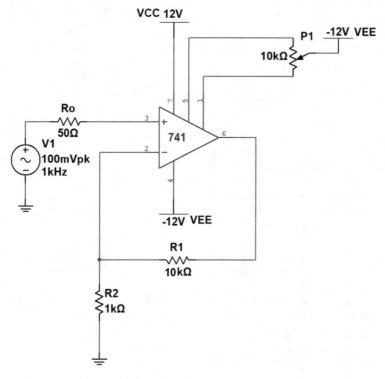

Fig. 6.12 Input is applied to the amplifier after nulling the offset

The procedure described here is not limited to the non-inverting amplifier. It can be used for other types of amplifiers as well.

6.5   Inverting Amplifier

An inverting amplifier, i.e. there is a 180° phase difference between input and output of the amplifier, is shown in Fig. 6.13. Gain of this amplifier is $A_v = -\frac{R_2}{R_1}$.

Fig. 6.13 Inverting amplifier

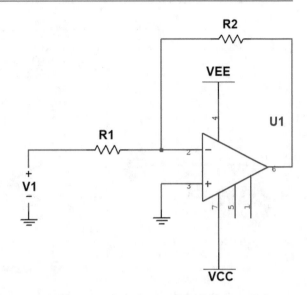

Make the circuit shown in Fig. 6.14. RO shows the output resistance of signal generator V1. Measure the phase difference between input and output. Measure the voltage gain (i.e. peak-to-peak of output voltage divided by peak-to-peak of input) of the amplifier as well. Compare the measured value with the value predicted by theory, i.e., $A_v = -\frac{R_2}{R_1} = -10$.

Observe the input and output simultaneously. Note that presence of negative feedback reduces the nonlinear effects and force the amplifier to behave linearly, i.e., input is amplified with almost constant gain everywhere.

Increase the input amplitude and measure the maximum input amplitude that cause no clipping in the output.

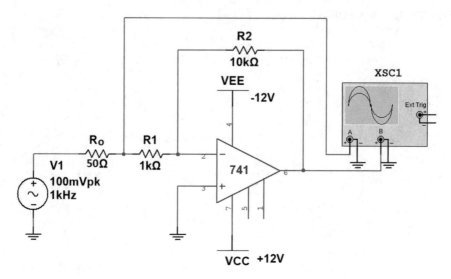

Fig. 6.14 Inverting amplifier used in this experiment

If you need to neutralize the output offset, you can use the circuit shown in Fig. 6.15.

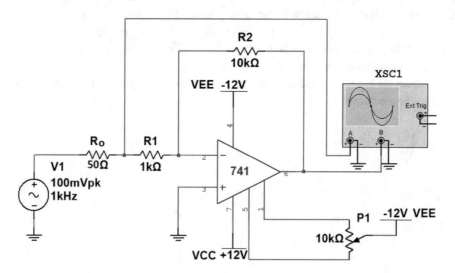

Fig. 6.15 Potentiometer P1 nulls the offset

6.6 Buffer

A buffer circuit (Fig. 6.16) prevents the signal source to be affected by the load. Buffer circuit is a current amplifier with voltage gain of around unity (i.e. $V_{out} = V_{in}$).

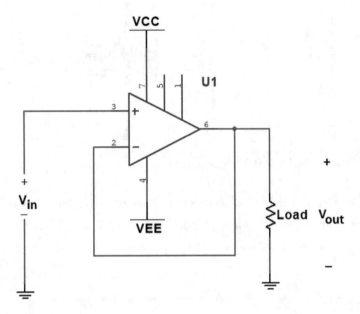

Fig. 6.16 Buffer circuit

Let's study an example. Consider the simple circuit shown in Fig. 6.17. Voltage of resistor R2 is $\frac{1k}{1k+1k} \times 20 = 10\,V$.

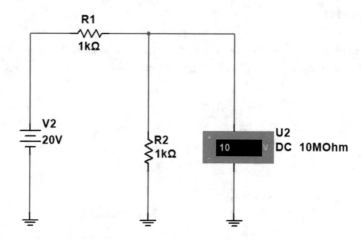

Fig. 6.17 Simple voltage divider circuit

Now a load with value of 1 kΩ is connected in parallel to resistor R2 (Fig. 6.18). Note that voltage is decreased to $\frac{\frac{1k \times 1k}{1k+1k}}{\frac{1k \times 1k}{1k+1k}+1} \times 20 = \frac{0.5k}{0.5k+1k} \times 20 = 6.667\,\text{V}$. In other words, added resistor loaded the circuit.

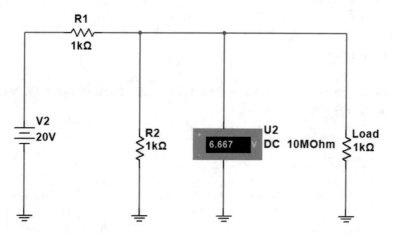

Fig. 6.18 Voltage divider circuit is loaded

Change the circuit to what shown in Fig. 6.19. As you see the voltage of load resistor is 10 V in this case. In other words, the voltage divider section (Voltage source V2 and resistors R1 and R2) is not loaded in this circuit.

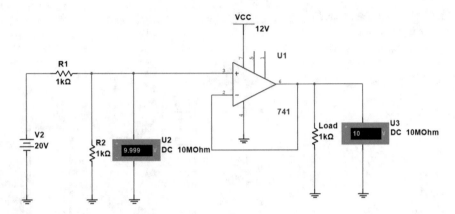

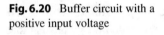

Fig. 6.19 Buffer avoids loading effect

Make the circuit shown in Fig. 6.20. Use the potentiometer P1 to set the potential of node P equal to 2, 4, 6 and 8 V. Measure the output for each voltage. Compare the output voltage with the associated input voltage. Are they equal?

Fig. 6.20 Buffer circuit with a positive input voltage

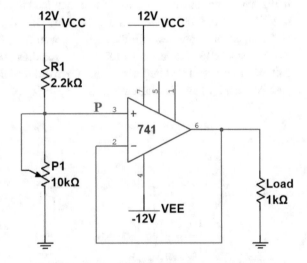

Change the circuit to what shown in Fig. 6.21. Use the potentiometer P1 to set the potential of node P equal to −2, −4, −6 and −8 V. Measure the output for each voltage. Compare the output voltage with the associated input voltage. Are they equal?

Fig. 6.21 Buffer circuit with a
negative input voltage

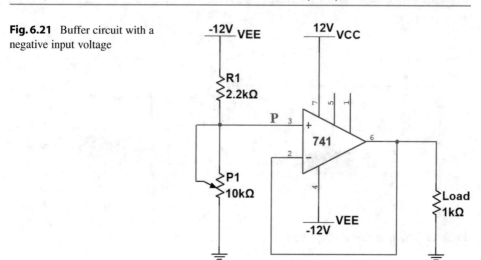

6.7   Improving the Current Driving Capability of a Buffer

In the schematic shown in Fig. 6.22 the output load is 100 Ω. However, the output voltage
is not 10 V, it is 2.53 V. Note that output of the Op Amp can supply a limited current and
buffering action continues until the current drawn from the output doesn't go beyond that
value. Keeping 10 V across a 100 Ω resistor requires 100 mA and the 741 cannot supply
100 mA. Because of that buffering is not observed for 100 Ω load. This experiment shows
how you can improve the current driving capability of your buffer.

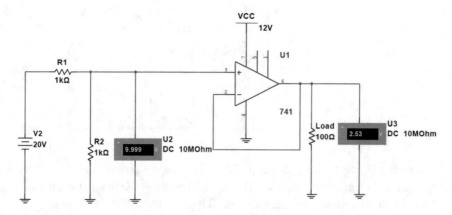

Fig. 6.22 Buffering is not observed

Let's get started. Make the circuit shown in Fig. 6.23. Set the voltage of node P equal
to 10 V and measure the voltage of node out. Do you observe the 10 V in the output?

Fig. 6.23 Buffer circuit with a 100 Ω load

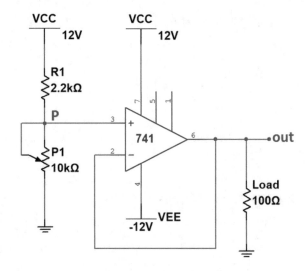

Change the circuit to what shown in Fig. 6.24 (Use a 100 Ω 1 W resistor). Pinout of 2N3904 and 2N2904 are shown in Figs. 6.25 and 6.26, respectively. Set the voltage of node P equal to 10 V. Do you observe the 10 V in output? Which one of the transistors handles the load current in this case?

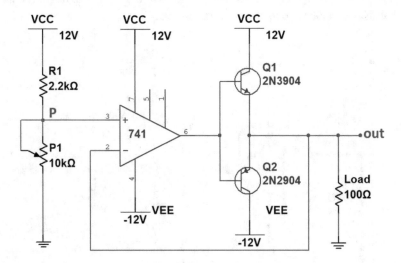

Fig. 6.24 Buffer circuit with a positive input voltage

Fig. 6.25 Pinout of 2N3904

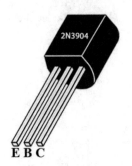

Fig. 6.26 Pinout of 2N2904

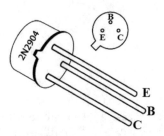

Now change the circuit to what shown in Fig. 6.27 (Use a 100 Ω 1 W resistor). Set the voltage of node P equal to −10 V. Do you observe the −10 V in output? Which one of the transistors handles the load current in this case?

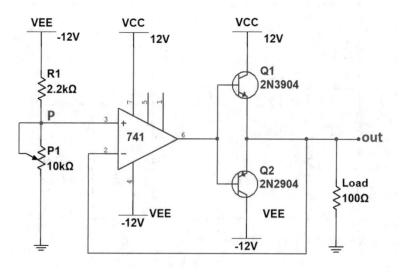

Fig. 6.27 Buffer circuit with a negative input voltage

Explain how circuit shown in Figs. 6.24 or 6.27 works?

6.8   Difference Amplifier

A difference amplifier is shown in Fig. 6.28. Output voltage for this circuit equals to $V_{out} = \frac{R_2}{R_1}(V_2 - V_1)$.

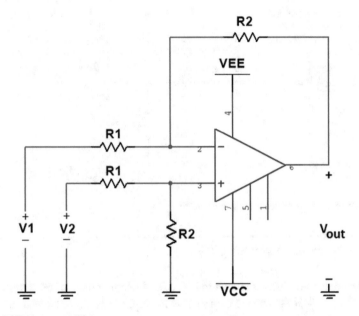

Fig. 6.28 Difference amplifier

Make the circuit shown in Fig. 6.29. Use a DC voltmeter to set the voltage of node V2 equal to 0.5 V. Measure the voltage of node out. Do you observe $\frac{R_2}{R_1}(V_2 - V_1) = \frac{10k}{1k}(0.5 - 0) = 5\,V$ at node out?

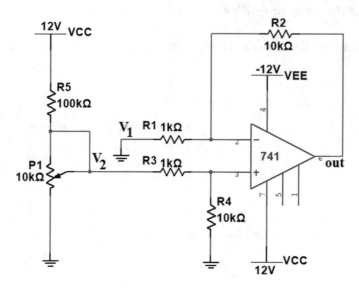

Fig. 6.29 Difference amplifier used in the circuit

6.9   Instrumentation Amplifier

An instrumentation amplifier is shown in Fig. 6.30. Output voltage of this amplifier equals to $V_{out} = \frac{R_4}{R_3}(1 + \frac{R_2}{R_1})(V_2 - V_1)$. Instrumentation amplifier is a more advanced difference amplifier. The input impedance of the instrumentation amplifier shown in Fig. 6.30 is higher than the difference amplifier shown in Fig. 6.28. From theoretical point of view, the current drawn from signal source V1 and V2 in Fig. 6.30 is 0 because they are connected to + and − input terminals of the Op Amps and from theoretical point of view, no current is drawn by input terminals of an ideal Op Amp. So, signal sources V1 and V2 in Fig. 6.30 see input resistance of infinity.

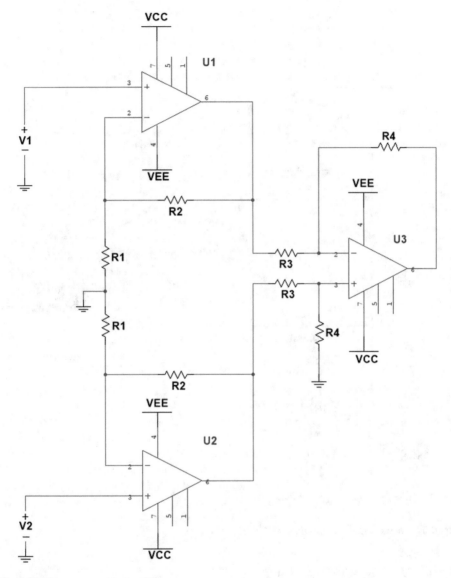

Fig. 6.30 Instrumentation amplifier

Let's get started. Make the circuit shown in Fig. 6.31. Use a DC voltmeter to set the voltage of node V2 equal to 0.5 V. Measure the voltage of node out. Do you observe $\frac{R_4}{R_3}\left(1 + \frac{R_2}{R_1}\right)(V_2 - V_1) = \frac{4.7k}{3.3k}\left(1 + \frac{2.2k}{1k}\right)(0.5 - 0) = 2.28\,\text{V}$ at node out?

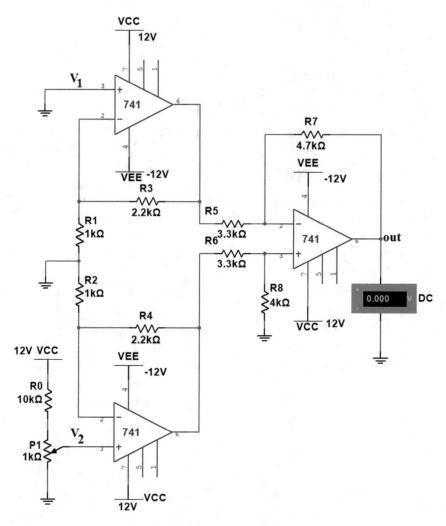

Fig. 6.31 Instrumentation amplifier used in this experiment

6.10 Voltage Comparator

The Op Amp comparator compares one analogue voltage level with another analogue voltage level and produces an output signal based on this voltage comparison. In other words, the Op Amp voltage comparator compares the magnitudes of two voltage inputs and determines which one is the largest of the two. Different types of comparator circuits are studied in this and next three experiments.

Let's get started. Make the circuit shown in Fig. 6.32. In this circuit voltage of node N is constant and equals to 4.8 V. Connect a DC voltmeter to node P and use the potentiometer P1 to set the voltage of node P equals to +3 V. Measure the voltage of node out. Now set the voltage of node P equal to +5 V. Measure the voltage of node out again. Explain how this circuit works.

Fig. 6.32 Simple voltage comparator

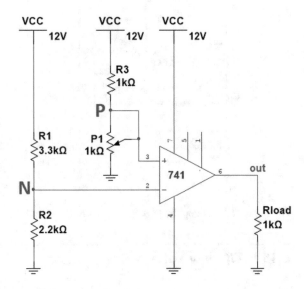

Use the potentiometer P1 to set the voltage of node P to minimum, i.e., around zero. Connect a DC voltmeter to the output in order to monitor output voltage (Fig. 6.33). Output voltage is around 0 V for this case. Now slowly increases the potential of node P and try to determine the voltage of node P which cause the output jumps to high level, i.e. around 11 V.

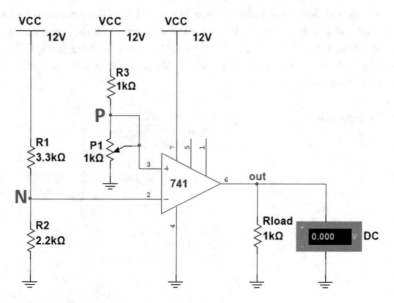

Fig. 6.33 Determining the triggering point of the circuit

6.11   Window Voltage Comparator

Window comparators are used to determine whether an unknown input voltage is between two reference voltages (V_{bot} and V_{top} in Fig. 6.34). Output Y of Fig. 6.34 is high when $V_{bot} < V_{in} < V_{top}$. Output is low when $V_{bot} < V_{in} < V_{top}$ is not satisfied.

Window comparators are used to detect over-voltage or under-voltage conditions. Window comparators are studied in this experiment.

Fig. 6.34 Block diagram of a window comparator

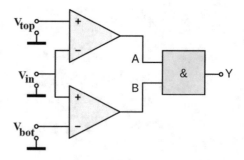

Let's get started. Make the window comparator circuit shown in Fig. 6.35. Note that supply circuit for 7408 IC is not shown in Fig. 6.35, it is shown in Fig. 6.36.

In Fig. 6.35, voltage of nodes P1 and N2 are constant and around 2.79 V and 0.92 V, respectively. Voltage of node out is high, i.e., around 5 V, when input voltage is bigger than 0.92 V and less than 2.79 V. When input voltage is outside of [0.92, 2.79 V] interval, voltage of node out is low, i.e., around 0 V.

Use the potentiometer P1 to set the voltage of V_{in} around zero. Output (node out) is low, i.e., around 0 V, for this case. Increase the voltage V_{in} slowly. Determine the input voltage V_{in} which cause a jump from low to high in output. Continue increasing the V_{in} and determine the V_{in} which cause a jump from high to low in output. Compare the measured values with the values predicted by theory (i.e., 0.92 and 2.79 V).

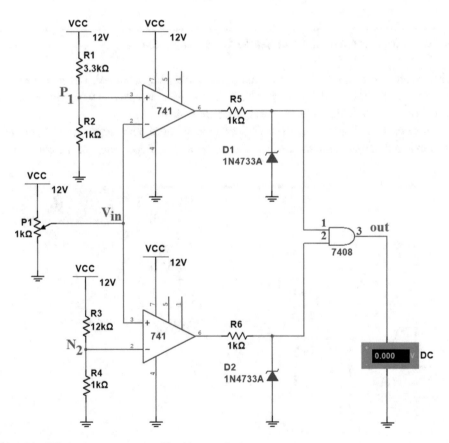

Fig. 6.35 Window comparator used in this experiment

Fig. 6.36 Supply circuit for
7408 IC used in Fig. 6.35

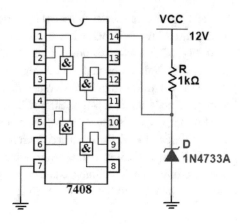

6.12   Conversion of Sine Wave to Square Wave

The circuit shown in Fig. 6.37 compares a sine wave with zero level. When sine wave is
positive, the output is around +5 V. When the sine wave is negative the output is around
0 V. Note that the IC is supplied with symmetric voltages.

Make the circuit shown in Fig. 6.37 (set the coupling of both channel to DC). Observe
the input and output simultaneously. Measure the high and low levels of the output.

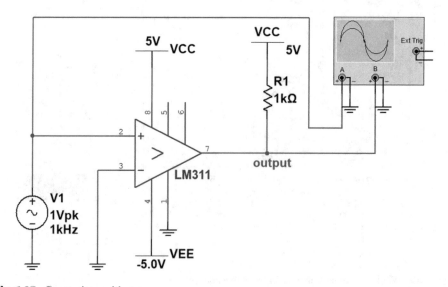

Fig. 6.37 Comparison with zero

Now make the circuit shown in Fig. 6.38 (set the coupling of both channel to DC). Observe the input and output simultaneously. Measure the high and low levels of the output.

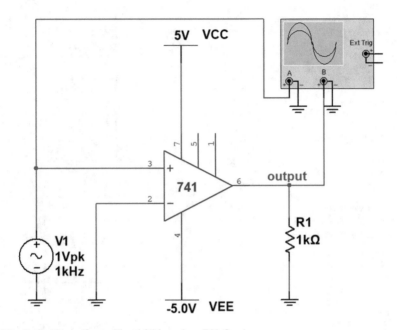

Fig. 6.38 A circuit similar to Fig. 6.37 based on 741 Op Amp

Compare the output waveforms of circuits shown in Figs. 6.37 and 6.38 with each other. Output of which one of the circuits is TTL compatible? Which one goes quicker from one output level to another?

6.13 Schmitt Trigger

A Schmitt trigger is a comparator with hysteresis. A simple Schmitt trigger circuit and its transfer function are shown in Figs. 6.39 and 6.40, respectively. In Fig. 6.40, M shows the saturation voltage of the Op Amp and $T = \frac{R1}{R2}M$ shows the trigger points. Note that Fig. 6.39 uses positive feedback.

Fig. 6.39 Schmitt trigger
circuit

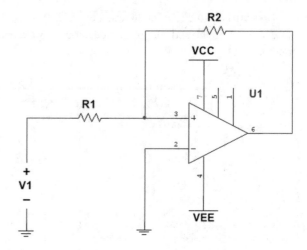

Fig. 6.40 Input-output
characteristic of Fig. 6.39 (M
shows the saturation voltage of
the Op Amp and
$T = \frac{R1}{R2} \times M$)

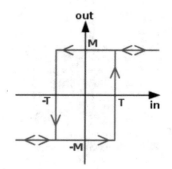

Make the circuit shown in Fig. 6.41. RO shows the output resistance of signal generator V1. Signal generator V1 generates a triangular voltage with peak value of 5 V and
frequency of 100 Hz (Fig. 6.42). Observe the input and output waveforms simultaneously.
Measure the trigger points and compare them with the values of predicted by theory.

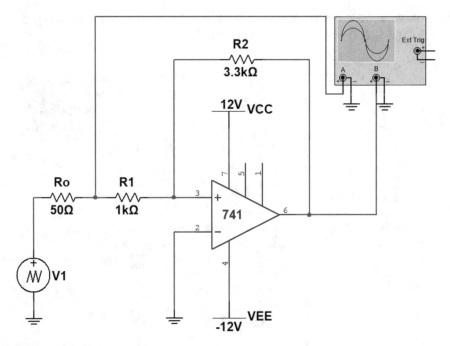

Fig. 6.41 Schmitt trigger circuit used in the experiment

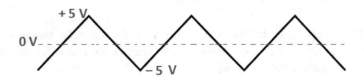

Fig. 6.42 Waveform of input voltage source V1

Triggering points of circuit shown in Fig. 6.39 depends on the saturation voltage of the Op Amp. Saturation voltage of an Op Amp is not a constant value and changes from Op Amp to Op Amp. Another problem is that positive and negative saturation voltages may not be equal in magnitude, for instance positive saturation may occur at $+1$ V and negative saturation may occur at -11.2 V. This means that trigger levels are not symmetric.

Circuit shown in Fig. 6.43 solves aforementioned problems. In this circuit voltage of node out (which is feed backed to positive terminal of the Op Amp) is determined by voltage of Zener diodes. When pin 6 of the Op Amp is saturated and its voltage is close to VCC, D1 is reverse biased and D2 is forward biased. Therefore, voltage of node out is around $5.1 + 0.7 = 5.8$ V. Another case is when pin 6 of Op Amp is saturated and its voltage is close to VEE. In this case D1 is forward biased and D2 is reverse biased and voltage of node out is around $-5.1-0.7 = -5.8$ V. Trigger levels of Fig. 6.43 is $T = \pm\frac{R1}{R2}(V_z + 0.7) = \pm\frac{1k}{3.3k}(5.1 + 0.7) = \pm1.76$ V.

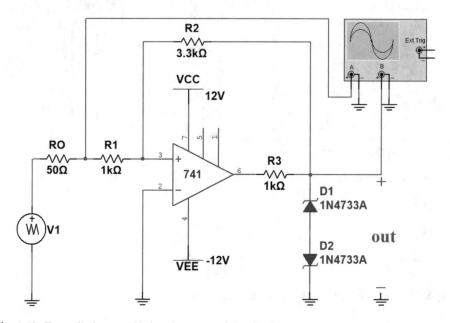

Fig. 6.43 Zener diodes are added to the output of the circuit

Make the circuit shown in Fig. 6.43. Apply an input waveform similar to Fig. 6.42 and observe the input and output simultaneously. Measure the trigger levels and compare them with the values predicted by theory.

6.14   Slew Rate

In electronics, the slew rate is defined as the maximum rate of output voltage change per unit time. Let's measure the slew rate of the 741 Op Amp. Make the circuit shown in Fig. 6.44. RO shows the output resistance of the signal generator V1. Signal generator V1 generates a square wave with frequency of 1 kHz and 0 and 5 V levels.

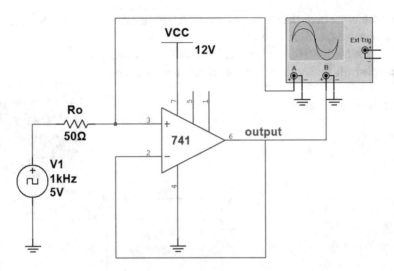

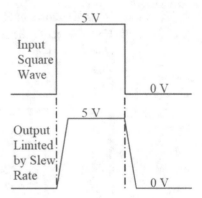

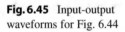

Fig. 6.44   Circuit to measure the slew rate of the Op Amp

Observe the input and output simultaneously. As shown in Fig. 6.45, the output can't change instantaneously and it requires some time to go from one level to another. Measure the slope of rising/falling edge of the output. The measured slope gives the slew rate of the Op Amp to you.

Fig. 6.45   Input-output
waveforms for Fig. 6.44

6.15   Half Wave Precision Rectifier

The precision rectifier is a configuration obtained with an operational amplifier in order to have a circuit behave like an ideal diode, i.e., a diode with zero voltage drop, and rectifier.

It is very useful for high-precision signal processing. This and next two experiment studies different types of precision rectifiers.

Let's get started. Make the circuit shown in Fig. 6.46. Observe the input and output simultaneously. Is it able to (half wave) rectify the input signal?

Fig. 6.46 Half wave rectifier

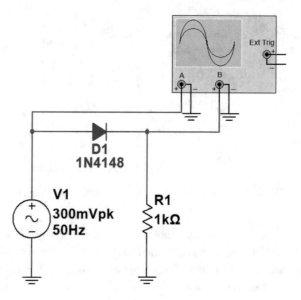

Now make the circuit shown in Fig. 6.47. Observe the input and output simultaneously. Is it able to (half wave) rectify the input signal?

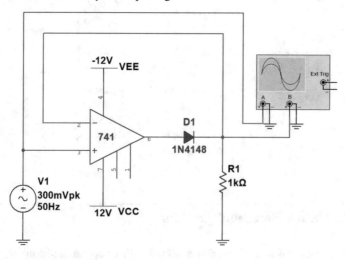

Fig. 6.47 Half wave precision rectifier

Change the direction of the diode D1 (Fig. 6.48). What happens to output waveform?

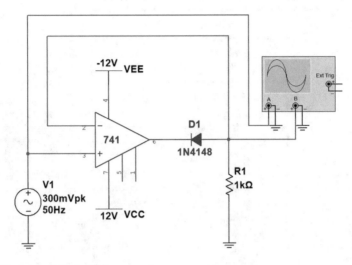

Fig. 6.48   Direction of diode D1 is reversed

6.16   Improved Half Wave Rectifier

Figure 6.49 shows an improved half wave precision rectifier circuit. This circuit provides a gain equal to $\frac{R_2}{R_1}$ as well.

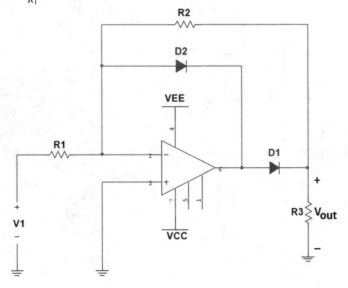

Fig. 6.49   Improved half wave precision rectifier circuit

Let's get started. Make the circuit shown in Fig. 6.50. Observe the input and output simultaneously. Compare the peak of output with the input.

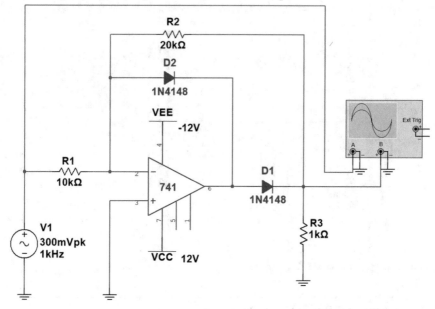

Fig. 6.50 Circuit used in this experiment

Now change the direction of diodes (Fig. 6.51). What happens to the output?

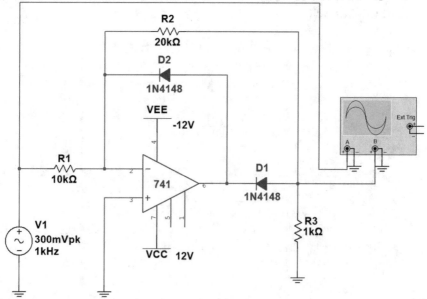

Fig. 6.51 Direction of diode D1 and D2 are reversed

6.17   Full Wave Rectifier

A full wave precision rectifier circuit is shown in Fig. 6.52. In this circuit $V_{out} = |V1|$.

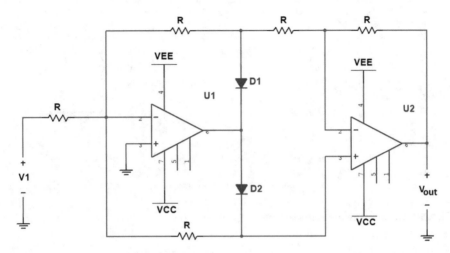

Fig. 6.52 Full wave precision rectifier circuit

You can change the direction of diodes (Fig. 6.53) and obtain a circuit with $V_{out} = -|V1|$.

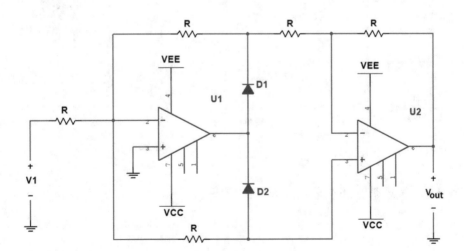

Fig. 6.53 Direction of diodes D1 and D2 are reversed

Let's get started. Make the circuit shown in Fig. 6.54. Observe the input and output simultaneously. Is the output full wave rectified version of input?

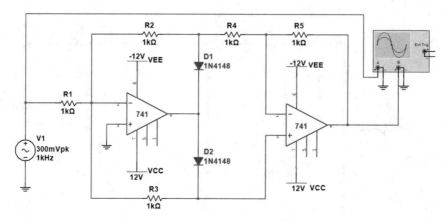

Fig. 6.54 Circuit used in this experiment

Now change the direction of the diodes (Fig. 6.55). What happens to the output waveform?

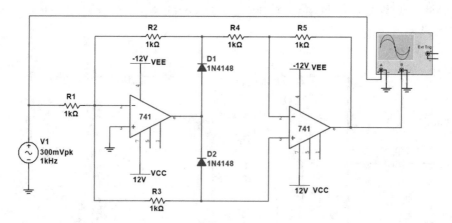

Fig. 6.55 Direction of diodes D1 and D2 are reversed

6.18   Low Pass Filter

A first order low pass filter is shown in Fig. 6.56. Transfer function of this circuit is $H(s) = (1 + \frac{R_1}{R_2}) \frac{1}{RCs+1}$. Pass band gain of this circuit is $(1 + \frac{R_1}{R_2})$. Cut-off frequency of this filter is $f_c = \frac{1}{2\pi RC} Hz$ or $\omega_c = \frac{1}{RC} \frac{Rad}{s}$.

Fig. 6.56 Low pass filter

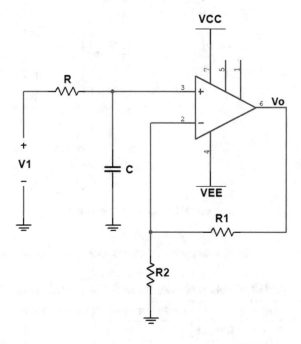

Let's get started. Make the circuit shown in Fig. 6.57. RO shows the output resistance of the signal generator V1. Transfer function of this circuit is $H(s) = \left(1 + \frac{R_1}{R_2}\right)\frac{1}{RCs+1} = 2.83 \times \frac{1}{0.291ms+1} = \frac{2.83}{\frac{s}{3.432k}+1}$.

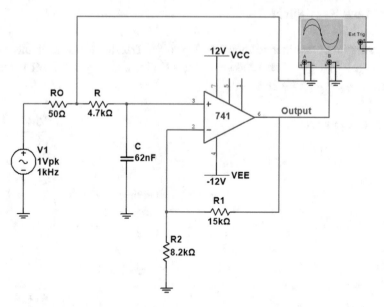

Fig. 6.57 Low pass filter used in this experiment

Set the frequency to the values shown in Table 6.1 and fill it.

Table 6.1 Gain of the filter shown in Fig. 6.57 for different frequencies

Frequency	100 Hz	200 Hz	300 Hz	550 Hz	600 Hz	800 Hz	1 kHz	5 kHz	10 kHz
Vp − p for Channel B ($V_{pp,B}$)									
Vp − p for Channel A ($V_{pp,A}$)									
Gain ($\frac{V_{pp,B}}{V_{pp,A}}$)									
Gain in dB ($20 \times \log_{10}(\frac{V_{pp,B}}{V_{pp,A}})$)									

Use MATLAB to draw the graph of Table 6.1. Let's measure the −3 dB cut-off frequency of the filter. In order to do this, set the frequency to a low value (frequency must be low enough to ensure that you are in the pass band). In this example you can set the frequency to 50 Hz. Now, measure the amplitude of the output. Multiply the measured amplitude by 0.707. Let's call the obtained value target amplitude. Don't change the amplitude of input signal and increase the frequency slowly. This cause the amplitude of the output start to decrease. The frequency at which the output amplitude reaches the target amplitude is the cut-off frequency.

Graph of $H(s) = \frac{2.83}{\frac{s}{3.432k}+1}$ can be drawn with the aid of commands shown in Fig. 6.58. Output of this code is shown in Fig. 6.59.

```
Command Window
>> options = bodeoptions;
>> options.FreqUnits = 'Hz';
>> s=tf('s');
>> H=2.83/(s/3.432e3+1);
>> bode(H,options)
>> grid on
fx >>
```

Fig. 6.58 MATLAB commands

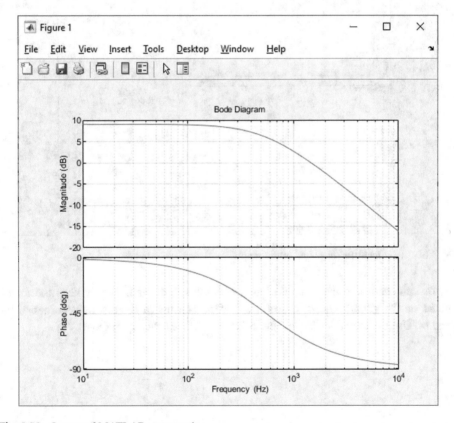

Fig. 6.59 Output of MATLAB commands

You can read different points of the graph by clicking on them (Fig. 6.60). According to Fig. 6.60, the pass band gain is around 9.03 dB. This means the gain in the pass band is $10^{\frac{9.03}{20}} = 2.83$. The -3 dB frequency is around 547 Hz and slope of transitions from pass band to stop band is around $\frac{-16.2dB - 2.63dB}{1 decade} = -18.83 \frac{dB}{decade}$.

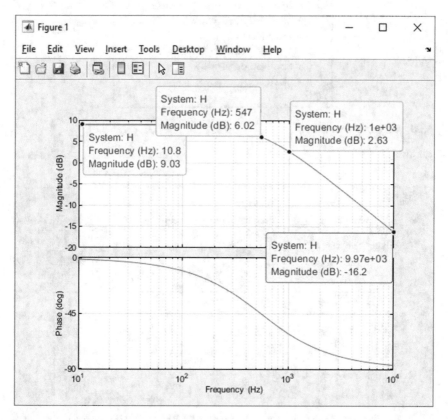

Fig. 6.60 Reading different points of the graph

You can zoom into the desired section of the graph as well. Simply keep the mouse pointer on the graph. After few seconds a menu appears in the top right corner of the window. Use the magnifier icons (Fig. 6.61) to select the desired region.

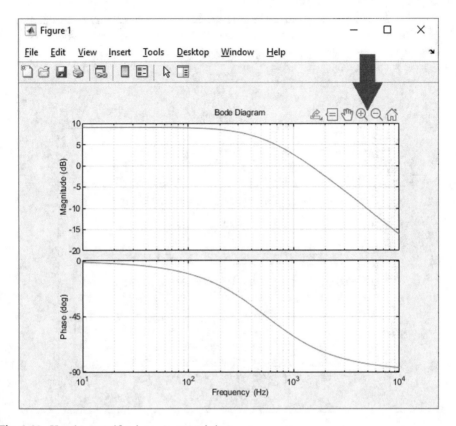

Fig. 6.61 Use the magnifier icons to zoom in/out

The code shown in Fig. 6.62 shows how to draw the frequency response in a given range. The code shown in Fig. 6.62 draws the frequency response for [100 Hz, 7.5 kHz] interval. Output of this code is shown in Fig. 6.63.

Fig. 6.62 MATLAB commands

```
Command Window
>> options = bodeoptions;
>> options.FreqUnits = 'Hz';
>> s=tf('s');
>> H=2.83/(s/3.432e3+1);
>> fmin=100;fmax=7500;
>> w=logspace(log10(2*pi*fmin),log10(2*pi*fmax));
>> bode(H,w,options)
>> grid on
>>
fx
```

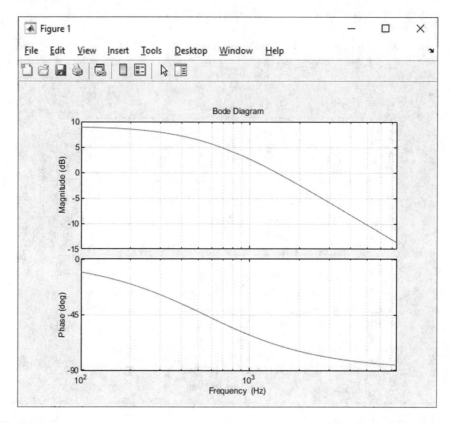

Fig. 6.63 Output of MATLAB commands

Compare the measured values (Table 6.1) with the corresponding points of graph shown in Fig. 6.59. Try to explain the reason of any discrepancy.

6.19   Cascaded Filters

In Fig. 6.64 two low pass filters are cascaded with each other. Such a cascade connection provides a steeper transition from pass band to stop band. Transfer function of circuit shown in Fig. 6.64 is $H(s) = \left(1 + \frac{R_3}{R_2}\right) \frac{1}{R_1 C_1 s + 1} \times \left(1 + \frac{R_6}{R_5}\right) \frac{1}{R_4 C_2 s + 1} = 2.83 \frac{1}{0.291 ms + 1} \times$ $2.83 \frac{1}{0.291 ms + 1} = \frac{8}{(0.291 ms + 1)^2} = \frac{9.447 \times 10^7}{(s + 3.44 k)^2} = \frac{7.98}{(1 + \frac{s}{3.44 k})^2}.$

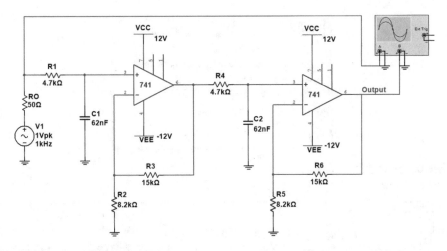

Fig. 6.64 Cascaded low pass filter

Let's get started. Make the circuit shown in Fig. 6.64. RO shows the output resistance of the signal generator V1. Set the frequency to the values shown in Table 6.2 and fill it.

Table 6.2 Gain of the filter shown in Fig. 6.64 for different frequencies

Frequency	100 Hz	200 Hz	300 Hz	550 Hz	600 Hz	800 Hz	1 kHz	5 kHz	20 kHz
Vp − p for Channel B $(V_{pp,B})$									
Vp − p for Channel A $(V_{pp,A})$									
Gain $(\frac{V_{pp,B}}{V_{pp,A}})$									

Use MATLAB to draw the graph of Tables 6.1 and 6.2 on the same axis. Compare them with each other. Which one has a steeper transition from passband to stop band? Measure the −3 dB cut-off frequency of the filter shown in Fig. 6.64 using the technique explained in Sect. 6.18.

Graph of $H(s) = \frac{7.98}{(1 + \frac{s}{3.44k})^2}$ can be drawn with the aid of commands shown in Fig. 6.65. Output of this code is shown in Fig. 6.66.

```
Command Window                            ⊙
  >> options = bodeoptions;
  >> options.FreqUnits = 'Hz';
  >> s=tf('s');
  >> H=7.98/(1+s/3.44e3)^2;
  >> bode(H,options)
  >> grid on
fx >>
```

Fig. 6.65 MATLAB commands

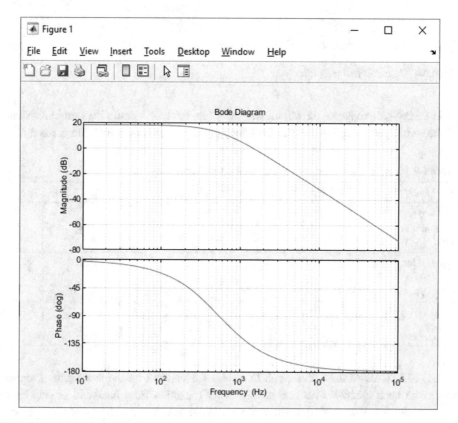

Fig. 6.66 Output of MATLAB commands

According to Fig. 6.67, the pass band gain is around 18 dB. This means the gain in the pass band is around $10^{\frac{18}{20}} = 7.94$. The -3 dB frequency is 355 Hz and slope of transitions from pass band to stop band is around $\frac{-33.1dB - 4.79dB}{1 decade} = -37.89 \frac{dB}{decade}$. This value is around 2 times bigger than the corresponding value for a single stage low pass filter.

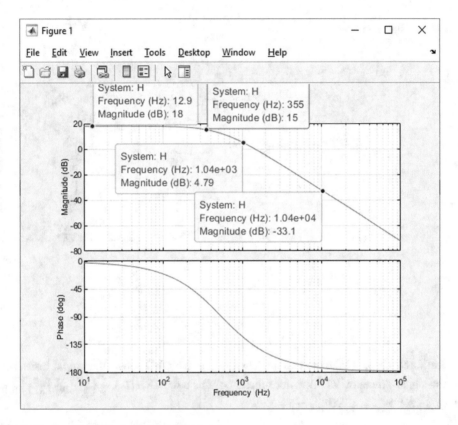

Fig. 6.67 Reading different points of the graph

Compare the measured values (Table 6.2) with the corresponding points of graph shown in Fig. 6.66. Try to explain the reason of any discrepancy.

6.20   High Pass Filter

A first order high pass filter is shown in Fig. 6.68. Transfer function of this circuit is $H(s) = (1 + \frac{R_1}{R_2})\frac{RCs}{RCs+1}$. Pass band gain of this circuit is $(1 + \frac{R_1}{R_2})$. Cut-off frequency of this filter is $f_c = \frac{1}{2\pi RC} Hz$ or $\omega_c = \frac{1}{RC}\frac{Rad}{s}$.

Fig. 6.68 High pass filter

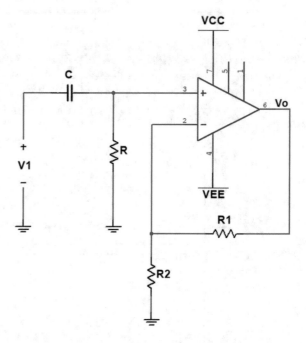

Let's get started. Make the circuit shown in Fig. 6.69. RO shows the output resistance of the signal generator V1. Transfer function of this circuit is $H(s) = \left(1 + \frac{R_1}{R_2}\right)\frac{RCs}{RCs+1} = 2.83 \times \frac{0.291ms}{0.291ms+1} = \frac{2.83s}{s+3.432k}$.

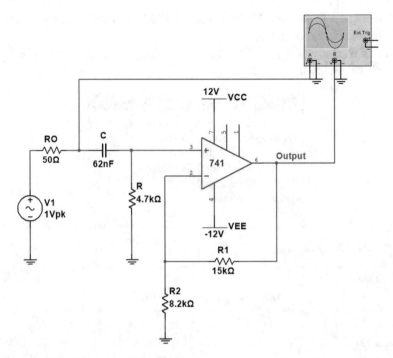

Fig. 6.69 High pass filter used in this experiment

Set the frequency to the values shown in Table 6.3 and fill it.

Table 6.3 Gain of the filter shown in Fig. 6.69 for different frequencies

Frequency	300 Hz	400 Hz	500 Hz	550 Hz	600 Hz	800 Hz	1 kHz	5 kHz	20 kHz
Vp − p for Channel B ($V_{pp,B}$)									
Vp − p for Channel A ($V_{pp,A}$)									
Gain ($\frac{V_{pp,B}}{V_{pp,A}}$)									
Gain in dB ($20 \times \log_{10}(\frac{V_{pp,B}}{V_{pp,A}})$)									

Use MATLAB to draw the graph of Table 6.3. Let's measure the −3 dB cut-off frequency of the filter. In order to do this, set the frequency to a high value (frequency must be high enough to ensure that you are in the pass band). In this example you can set the frequency to 5 kHz. Now, measure the amplitude of the output. Multiply the measured amplitude by 0.707. Let's call the obtained value target amplitude. Don't change the amplitude of input signal and decrease the frequency slowly. This cause the amplitude of the output start to decrease. The frequency at which the output amplitude reaches the target amplitude is the cut-off frequency.

Graph of $H(s) = \frac{2.83s}{s+3.432k}$ can be drawn with the aid of commands shown in Fig. 6.70. Output of this code is shown in Fig. 6.71.

```
Command Window
  >> options = bodeoptions;
  >> options.FreqUnits = 'Hz';
  >> s=tf('s');
  >> H=2.83*s/(s+3.432e3);
  >> bode(H,options)
  >> grid on
fx >>
```

Fig. 6.70 MATLAB commands

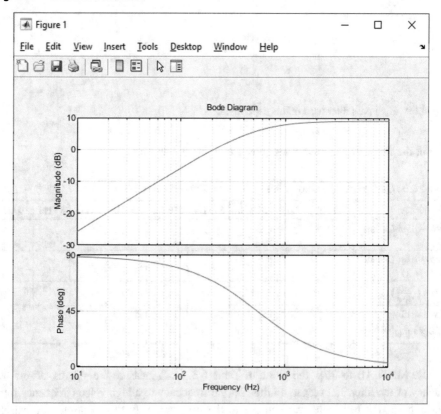

Fig. 6.71 Output of MATLAB commands

Compare the measured values (Table 6.3) with the corresponding points of graph shown in Fig. 6.71. Try to explain the reason of any discrepancy.

6.21   Band Pass Filter

A band pass filter is shown in Fig. 6.72. It is composed of two filters: A high pass filter and a low pass filter. Transfer function of the high pass section is $H_{hp}(s) = \left(1 + \frac{R_3}{R_2}\right)\frac{R_1C_1s}{R_1C_1s+1} = \frac{2.83s}{s+343.171}$. Transfer function of the low pass section is $H_{lp}(s) = \left(1 + \frac{R_6}{R_5}\right)\frac{1}{R_4C_2s+1} = \frac{2.83}{\frac{s}{3.43k}+1}$. Therefore, the overall transfer function is $H(s) = H_{hp}(s) \times H_{lp}(s) = \frac{2.83s}{s+343.171} \times \frac{2.83}{\frac{s}{3.43k}+1} = \frac{8s}{(s+343.171)(\frac{s}{3.43k}+1)}$.

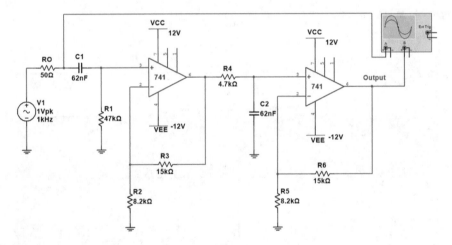

Fig. 6.72   Band pass filter used in this experiment

Let's get started. Make the circuit shown in Fig. 6.72. RO shows the output resistance of the signal generator V1. Set the frequency to the values shown in Table 6.4 and fill it.

Table 6.4   Gain of the filter shown in Fig. 6.72 for different frequencies

Frequency	30 Hz	50 Hz	100 Hz	150 Hz	200 Hz	300 Hz	500 Hz	650 Hz	800 Hz
$Vp - p$ for Channel B $(V_{pp,B})$									
$Vp - p$ for Channel A $(V_{pp,A})$									
Gain $(\frac{V_{pp,B}}{V_{pp,A}})$									
Gain in dB $(20 \times \log_{10}(\frac{V_{pp,B}}{V_{pp,A}}))$									

Use MATLAB to draw the graph of Table 6.4. Measure the low and high -3 dB cut-off frequency of the filter. Graph of $H(s) = \frac{8s}{(s+343.171)(\frac{s}{3.43k}+1)}$ can be drawn with the aid of commands shown in Fig. 6.73. Output of this code is shown in Fig. 6.74.

```
Command Window                                    ⊙

   >> options = bodeoptions;
   >> options.FreqUnits = 'Hz';
   >> s=tf('s');
   >> H=8*s/(s+343.171)/(s/3.43e3+1);
   >> bode(H,options)
   >> grid on
fx >>
```

Fig. 6.73 MATLAB code

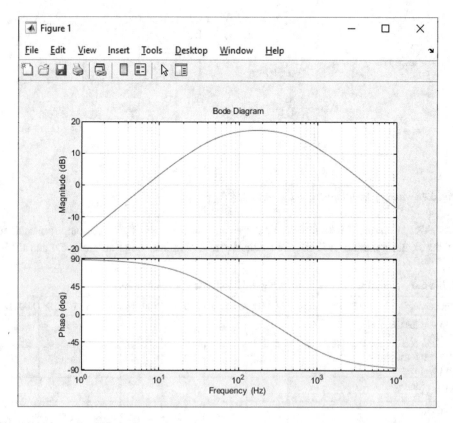

Fig. 6.74 Output of MATLAB code

Figure 6.75 shows the close up of pass band. According to Fig. 6.75 maximum gain of the filter is around 17.2 dB ($10^{\frac{17.2}{20}} = 7.24$). Low and high -3 dB frequencies are 46 Hz and 655 Hz, respectively.

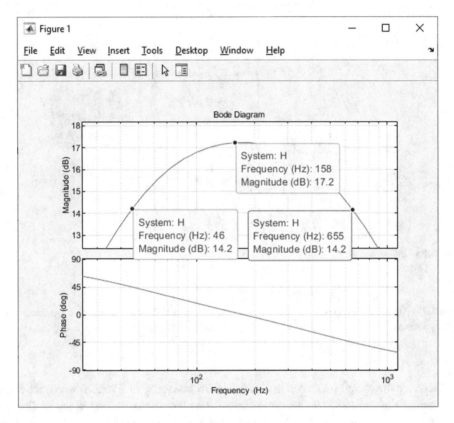

Fig. 6.75 Zoomed graph

Compare the measured values (Table 6.4) with the corresponding points of graph shown in Fig. 6.74. Try to explain the reason of any discrepancy.

6.22   Integrator Circuit

Op Amp integrators are studied in this experiment. Make the circuit shown in Fig. 6.76. RO shows the output resistance of signal generator V1. Signal generator V1 generates the signal shown in Fig. 6.77. Transfer function of Fig. 6.76 is $H(s) = -\dfrac{\frac{R_1}{R_2}}{R_1 C_1 s+1} = -\dfrac{10}{0.001s+1}$. When $0.001s \gg 1$ or equivalently $s \gg 1000$, $H(s) \approx -\dfrac{10^4}{s}$. Note that $s \gg$

1000 means $f \gg \frac{1000}{2\pi} = 159.15\,\text{Hz}$. Therefore, when $f \gg 159.15Hz$, output voltage is $V_{out}(t) = -10^4 \int V_1(t)dt$.

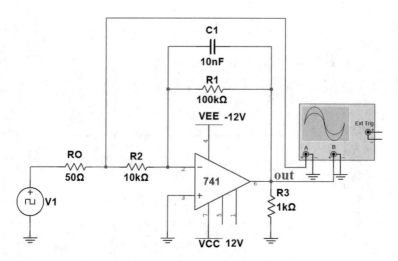

Fig. 6.76 Integrator circuit used in this experiment

Fig. 6.77 Waveform of
voltage source V1

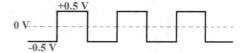

Generate the waveform shown in Fig. 6.77 with frequency of 2 kHz. Observe the input and output simultaneously. Is the output waveform close to the integral of input? Change the frequency of input signal to 200 Hz and repeat the experiment.

Give a sinusoidal signal with peak-to-peak value of 1 V, average of zero and frequency of 160 Hz to the circuit. Observe the input and output simultaneously. Measure the amplitude of output and phase difference between input and output. Compare the measured values with the value predicted using theory (**Hint**: Put $s = j2\pi \times 160 = 1000\frac{Rad}{s}$ in $-\frac{10}{0.001s+1}$ and calculate the magnitude and phase of obtained complex number).

6.23 Differentiator Circuit

Transfer function of circuit shown in Fig. 6.78 is $H(s) = -\frac{R_2C_1s}{R_1C_1s+1} = -\frac{10^{-5}s}{10^{-5}s+1}$. When $R_1C_1s \ll 1$, $H(s) \approx -R_2C_1s$. Note that $R_1C_1s \ll 1$ means that frequency of input signal satisfy $f \ll \frac{1}{2\pi R_1C_1}$. When $f \ll \frac{1}{2\pi R_1C_1} = 15.92kHz$, output is $-R_2C_1\frac{dV_1(t)}{dt}$.

Let's get started. Make the circuit shown in Fig. 6.78. Apply a square wave with frequency of 1 kHz (Fig. 6.79) to the circuit and observe the result. Increase the frequency to 10 kHz and observe the result.

Now give a sinusoidal signal with peak-to-peak value of 1 V, average of zero and frequency of 10 kHz to the circuit. Observe the input and output simultaneously. Measure the amplitude of output and phase difference between input and output. Compare the measured values with the value predicted using theory (**Hint**: Put $s = j2\pi \times 10k = 62.8\frac{kRad}{s}$ in $-\frac{10^{-5}s}{10^{-5}s+1}$ and calculate the magnitude and phase of obtained complex number).

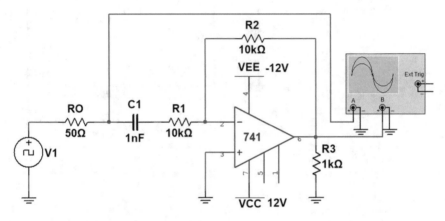

Fig. 6.78 Differentiator circuit used in this experiment

Fig. 6.79 Waveform of voltage source V1

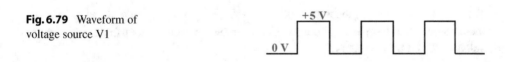

6.24 Wien-Bridge Oscillator

Schematic of a Wien-bridge oscillator is shown in Fig. 6.80. This circuit produces sinusoidal waveforms. Frequency of output waveform equals to $f = \frac{1}{2\pi RC}$. The Op Amp based Wien-bridge oscillator (with a suitable Op Amp) can be used to generate frequencies up to 1 MHz. $\frac{R_2}{R_1}$ must be a little bit bigger than 2 otherwise no oscillation is generated at the output.

Fig. 6.80 Wien bridge
oscillator

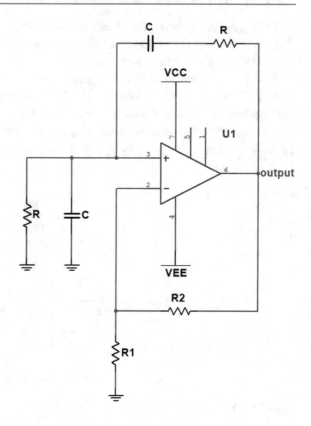

Let's get started. Make the circuit shown in Fig. 6.81. Note that $\frac{R_2}{R_1} = \frac{2.2k}{1k} = 2.2 > 2$. We expect an output frequency around $f = \frac{1}{2\pi RC} = \frac{1}{2\pi \times 10k \times 3.3n} = 4.823\,\text{kHz}$. Measure the amplitude and frequency of output. Compare the measured output frequency with the value predicted by the theory.

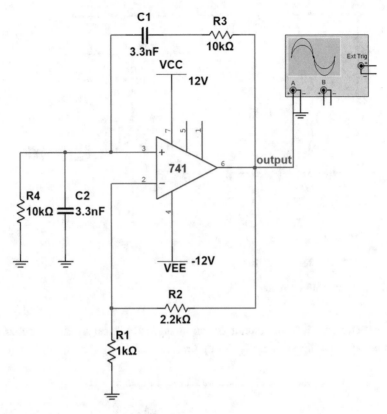

Fig. 6.81 Wien bridge oscillator used in this example

6.25   RC Phase Shift Oscillator

RC phase shift oscillator with Op Amp is shown in Fig. 6.82. Output frequency of this circuit equals to $f = \frac{1}{2\pi\sqrt{6}RC}$.

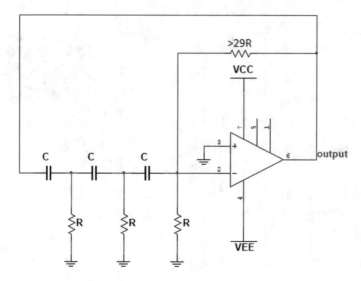

Fig. 6.82 RC phase shift oscillator

Let's get started. Make the circuit shown in Fig. 6.83. Measure the output frequency and compare it with the value predicted by theory.

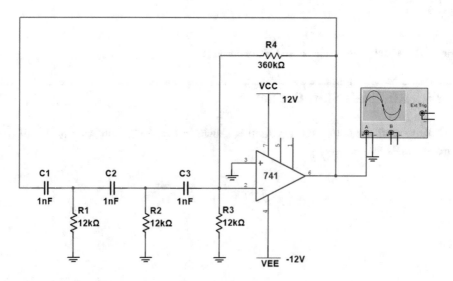

Fig. 6.83 RC phase shift oscillator used in this experiment

6.26 Square Wave Generator

A square wave generator with Op Amp is shown in Fig. 6.84. Output frequency for this circuit is $f = \frac{1}{2R_f C \ln(\frac{V_{sat}-V_{LT}}{V_{sat}-V_{UT}})}$. $V_{UT} = \frac{R_1+V_{sat}}{R_1+R_2}$ and $V_{LT} = \frac{R_1-V_{sat}}{R_1+R_2}$. Note that V_{sat} shows the saturation voltage of the Op Amp.

Fig. 6.84 Square wave generator

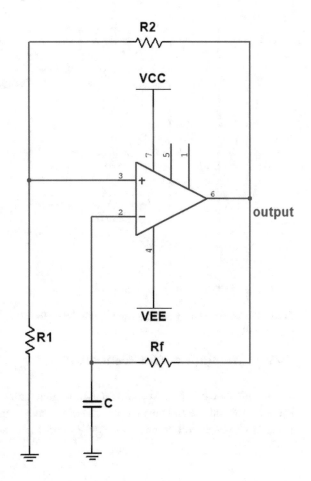

Let's get started. Make the circuit shown in Fig. 6.85. Measure the output frequency and compare it with the value predicted by theory.

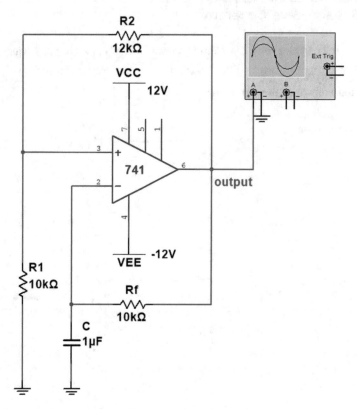

Fig. 6.85 Square wave generator used in this example

6.27 Triangular Wave Generator

The circuit shown in Fig. 6.86 can be used to generate triangular waveform. It generates a square wave and use an integrator to convert it into a triangle. Frequency and peak-to-peak value of output waveform are $f = \frac{R_0+P_1}{4R_1R_2C_1}$ and $V_{p-p} = \frac{2R_2}{R_0+P_1}V_{sat}$, respectively.

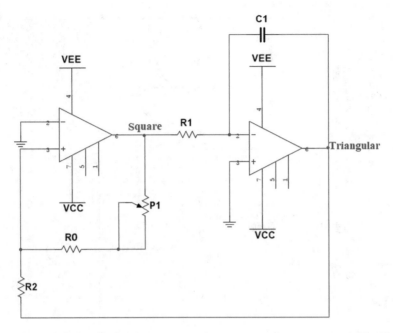

Fig. 6.86 Triangular wave generator

Let's get started. Make the circuit shown in Fig. 6.87. Set the $R_0 + P_1$ to be equal to 22 kΩ. Measure the output frequency and amplitude for this case. Compare the measured values with values predicted by theory.

Now turn the potentiometer P1 and measure the minimum and maximum of output frequency and amplitude.

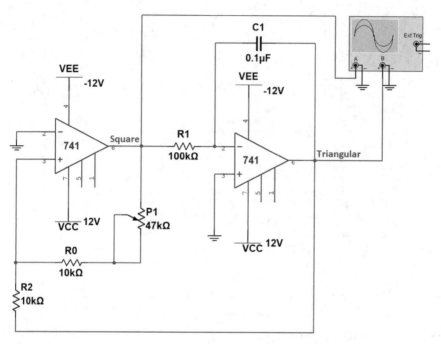

Fig. 6.87 Triangular wave generator circuit used in this experiment

6.28 Timer IC

The 555 timer IC (Fig. 6.88) is an IC used in a variety of timer, delay, pulse generation, and oscillator applications. Derivatives provide two (556) or four (558) timing circuits in one package. The design was first marketed in 1972 by Signetics. Since then, numerous companies have made the original bipolar timers, as well as similar low-power CMOS timers. In 2017, it was said that over a billion 555 timers are produced annually by some estimates, and that the design was "probably the most popular integrated circuit ever made".

Fig. 6.88 555 timer IC pinout

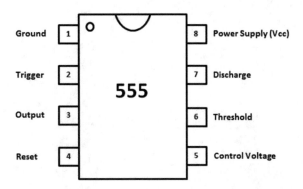

[1] is a very good reference on 555 timer IC circuits. Appendix B (Sect. B.2) introduced a software tool which designs the 555 circuits automatically. You can make the circuits shown in Sect. B.2 and see they work as expected.

References for Further Study

1. https://www.electronics-tutorials.ws/category/waveforms
2. Asadi F., Essential Circuit Analysis using Proteus, Springer, 2022. DOI: https://doi.org/10.1007/978-981-19-4353-9
3. Asadi F., Essential Circuit Analysis using LTspice, Springer, 2022. DOI: https://doi.org/10.1007/978-3-031-09853-6
4. Asadi F., Essential Circuit Analysis using NI Multisim and MATLAB, Springer, 2022. DOI: https://doi.org/10.1007/978-3-030-89850-2
5. Asadi F., Electric Circuit Analysis with EasyEDA, Springer, 2022. DOI: https://doi.org/10.1007/978-3-031-00292-2

Voltage Regulators and Power Amplifiers

7

7.1 Introduction

Voltage regulators provide a constant voltage for a circuit to work. Linear voltage regulators are studied in this chapter. In linear voltage regulators, the transistor is operated in the linear region. Power dissipated in the transistor decreases the efficiency of these type of regulators.

Power amplifiers are another important issue that is studied in this chapter. Power amplifiers are the final block in the amplifiers and they provide the required current to drive the load. Voltage gain of power amplifiers may be low however their current gain is high.

7.2 Series Voltage Regulator

A series voltage regulator is shown in Fig. 7.1 (Transistor Q1 is in series with the load resistor, that is why it is called series regulator). The 1N4733A Zener diode provides a 5.1 V at the positive terminal of the Op amp. The negative terminal of the Op Amp is connected to $\frac{3.3}{1+3.3} V_{Load} = 0.77 V_{Load}$. Presence of negative feedback force $0.77 V_{Load} = 5.1$. Therefore, output load voltage must be around 6.62 V.

© The Author(s), under exclusive license to Springer Nature Switzerland AG 2023
F. Asadi, *Analog Electronic Circuits Laboratory Manual*, Synthesis Lectures on Electrical Engineering, https://doi.org/10.1007/978-3-031-25122-1_7

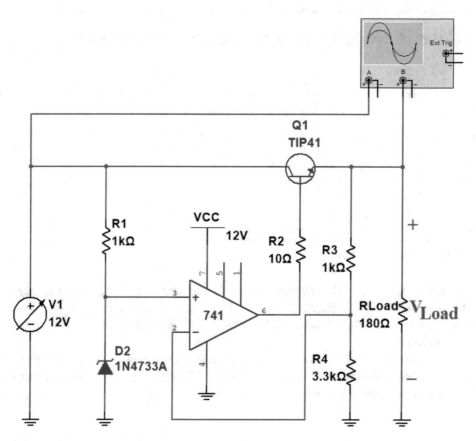

Fig. 7.1 Series voltage regulator circuit

Make the series voltage regulator shown in Fig. 7.1. Apply 12 V input voltage to the circuit and measure the load voltage. Change the input voltage to 15 and 9 V and re-measure the load voltage. Do you observe any considerable change in the load voltage?

7.3   Shunt Voltage Regulator

A shunt voltage regulator is shown in Fig. 7.2 (Transistor Q1 is in parallel with the load resistor, that is why it is called shunt regulator). The 1N4733A Zener diode provides a 5.1 V at the negative terminal of the Op amp. The positive terminal of the Op Amp is connected to $\frac{1}{1+1.5} V_{Load} = 0.4 V_{Load}$. Presence of negative feedback force $0.4 V_{Load} = 5.1$. Therefore, output load voltage must be around 12.75 V.

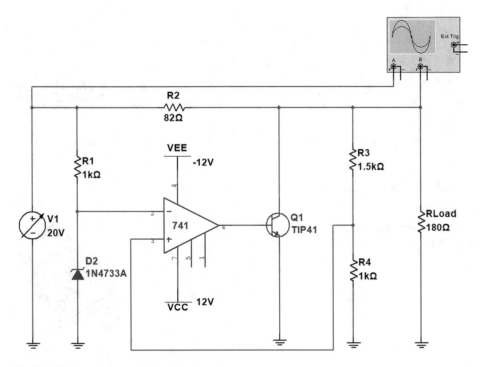

Fig. 7.2 Shunt voltage regulator circuit

Make the shunt voltage regulator shown in Fig. 7.2. Apply 20 V input voltage to the circuit and measure the load voltage. Change the input voltage to 24 and 16 V and re-measure the load voltage. Do you observe any considerable change in the load voltage?

7.4 Voltage Regulator IC's

78XX and 79XX IC's are among the most commonly used voltage regulators that used in the electronic circuits. 78XX family provides positive voltages while 79XX provides negative voltages. For instance, 7812 takes an input voltage V_i in the range $14\,V < V_i < 35\,V$ and gives $+12$ V output. 7912 takes an input voltage in the range $-14.5\,V < V_i < -27\,V$ and gives -12 V output. In this experiment we will use a 7805 to generate $+5$ V.

Make the circuit shown in Fig. 7.3. Pins of 7805 are shown in Fig. 7.4. 7805 works with voltages in the range of $10\,V < V_i < 35\,V$. Apply 12 V to the circuit and measure the output voltage. Increases the input voltage to 18 V and re-measure the output voltage. Do you observe considerable change in the output voltage?

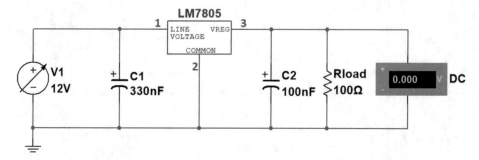

Fig. 7.3 7805 based voltage regulator circuit

Fig. 7.4 Pins of 7805 voltage
regulator

7.5   Class B Power Amplifier

In this experiment we will study a class B power amplifier. Make the circuit shown in
Fig. 7.5 (use a 10 Ω 1 W resistor as load). Pinout of TIP 41 and TIP 42 are shown in
Fig. 7.6. Set the amplitude of signal generator V1 to 3 V and observe the input and output
simultaneously (set the coupling of both channels to DC). Pay attention to the distortion.
Now increase the amplitude of input voltage to 10 V and repeat the experiment. In both
case pay attention to the distortion. Measure the input impedance R_{in} as well (use the
technique of Sect. 2.8 with Rx = R1 = 1 kΩ).

Fig. 7.5 Class B power
amplifier

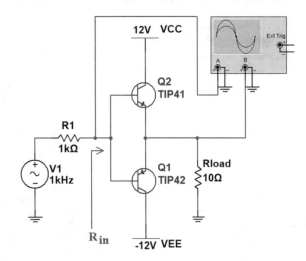

Fig. 7.6 Pinout of TIP 41 and
TIP 42

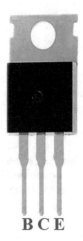

Let's measure the efficiency of the power amplifier shown in Fig. 7.5. Set the amplitude of input signal to 6 V. Use the oscilloscope to measure the RMS value of load's
voltage and call it $V_{load,RMS}$. The power consumed by the load resistor equals to
$P_{load} = \frac{V_{load,RMS}^2}{R_{load}}$. If you are using an analog oscilloscope, you can measure the
amplitude of load voltage ($V_{load,peak}$) and approximate the RMS value by using the
$V_{load,RMS} \approx \frac{V_{load,peak}}{\sqrt{2}}$ formula.

It is time to measure the power entered to the amplifier. Use a DC ammeter to measure
the current drawn from +12 V source (Fig. 7.7). The power drawn from +12 V source
equals to 12 times the current shown by the DC ammeter. Call the obtained number P1.

Fig. 7.7 Measurement of
average value (DC component)
of collector current

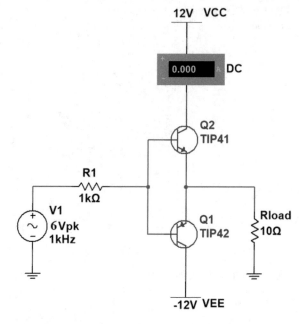

Now use the DC ammeter to measure the current drawn from −12 V source (Fig. 7.8). The power drawn from −12 V source equals to 12 times the current shown by the DC ammeter. Call the obtained number P2. Total power entered to the amplifier equals to the summation of P1 and P2.

Fig. 7.8 Measurement of
average value (DC component)
of emitter current

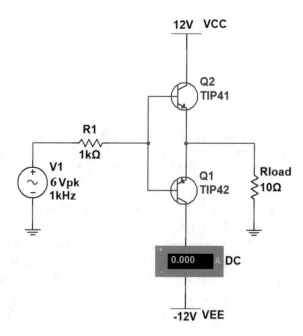

Efficiency of the amplifier can be calculated using the $\eta = \frac{P_{load}}{P_1+P_2} \times 100\%$ formula.

7.6   Class B Power Amplifier with Feedback

Distortion problem of previous experiment can be solved with the aid of feedback (Figs. 7.9 and 7.10). In Fig. 7.9 the input signal is applied to positive terminal of the Op Amp. Load voltage is connected to the negative terminal of the Op amp. Presence of the negative feedback force the output voltage to be equals to V1. Therefore, voltage gain of Fig. 7.9 is equal to 1.

In Fig. 7.10, input voltage is connected to the positive terminal of Op amp. $\frac{2.2}{2.2+2.2} V_{Load} = 0.5 V_{Load}$ is given to the negative terminal of the Op amp. Presence of negative feedback force the load voltage to be times the 2V1 since $0.5 V_{Load} = V_1 \Rightarrow V_{Load} = 2V_1$. Therefore, Fig. 7.10 provides a non-unity voltage gain as well.

Input resistance of circuits shown in Figs. 7.9 and 7.10 is very high since input source V1 is connected to the positive terminal of the Op amp. Remember that input terminals of an Op amp draw a very small current around zero. This gives a very high input resistance to the amplifier.

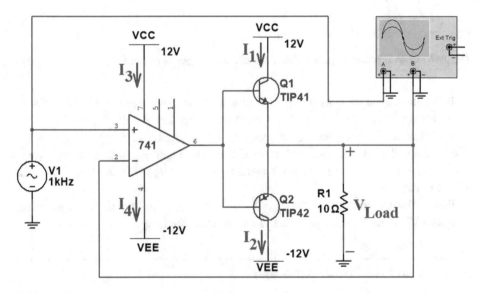

Fig. 7.9 Class B power amplifier with feedback

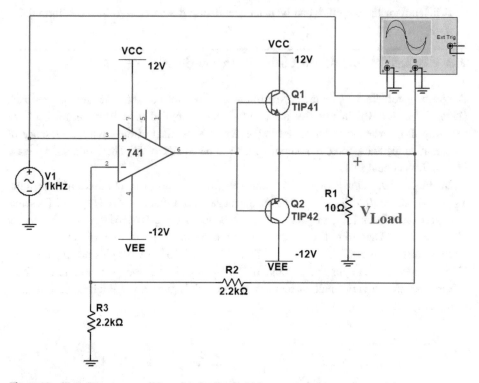

Fig. 7.10 Class B power amplifier with feedback. Voltage gain is bigger than unity

Make the circuit shown in Fig. 7.9 (use a 10 Ω 1 W resistor as load). Set the input voltage equals to 1 V and observe that output follows the input without distortion observed in previous experiment. Set the amplitude of input voltage to 3 V and measure the voltage gain and efficiency of the amplifier. Note that input power $P_{in} = V_{CC}(I_1 + I_2) + V_{EE}(I_3 + I_4) = 12 \times (I_1 + I_2 + I_3 + I_4)$. I_1, I_2, I_3 and I_4 are shown in Fig. 7.9.

Repeat this experiment for the circuit shown in Fig. 7.10. Set the peak of voltage generator V1 to 1.5 V when you want to measure the efficiency.

7.7 Class AB Power Amplifier

In this experiment class AB amplifier is studied. Make the circuit shown in Fig. 7.11 (use a 10 Ω 1 W resistor as load). Apply an input voltage with amplitude of 1 V and observe the output. Is there any distortion like the one you observed in Sect. 7.5? Measure the input resistance R_{in} of the amplifier as well (use the technique of Sect. 2.8 with Rx = R1 = 1 kΩ).

Apply an input voltage with amplitude of 5 V and measure the efficiency of the amplifier.

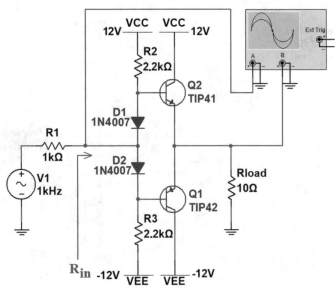

Fig. 7.11 Class AB power amplifier

Let's measure the current gain of the amplifier. Set the amplitude of V1 to 5 V and use an AC micro ammeter to measure the RMS of current drawn from source V1 (Fig. 7.12). Call the measured value $I_{in,RMS}$.

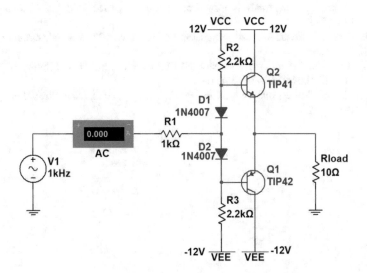

Fig. 7.12 Measurement of RMS value of current drawn from input voltage source V1

Now use the AC milli ammeter to measure the RMS of load current (Fig. 7.13). Call the measured value $I_{load,RMS}$.

Fig. 7.13 Measurement of RMS value of load current

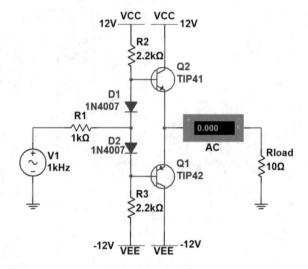

The current gain equals to $\frac{I_{load,RMS}}{I_{in,RMS}}$.

References for Further Study

1. Asadi F., Essential Circuit Analysis using Proteus, Springer, 2022. DOI: https://doi.org/10.1007/978-981-19-4353-9
2. Asadi F., Essential Circuit Analysis using LTspice, Springer, 2022. DOI: https://doi.org/10.1007/978-3-031-09853-6
3. Asadi F., Essential Circuit Analysis using NI Multisim and MATLAB, Springer, 2022. DOI: https://doi.org/10.1007/978-3-030-89850-2
4. Asadi F., Electric Circuit Analysis with EasyEDA, Springer, 2022. DOI: https://doi.org/10.1007/978-3-031-00292-2

Appendix A: Drawing Graphs with MATLAB®

A.1 Introduction

This appendix shows how to draw different types of graphs with MATLAB.

A.2 FPLOT Command

Assume that you want to draw the graph of $sin(x)$ for $[0, 2\pi]$ interval. The commands shown in Fig. A.1 do this job for you. Output of this code is shown in Fig. A.2.

```
Command Window
>> syms x
>> fplot(sin(x),[0,2*pi])
>> grid on
fx >>
```

Fig. A.1 FPLOT command can be used to draw symbolic expressions

F. Asadi, *Analog Electronic Circuits Laboratory Manual*, Synthesis Lectures on Electrical Engineering, https://doi.org/10.1007/978-3-031-25122-1

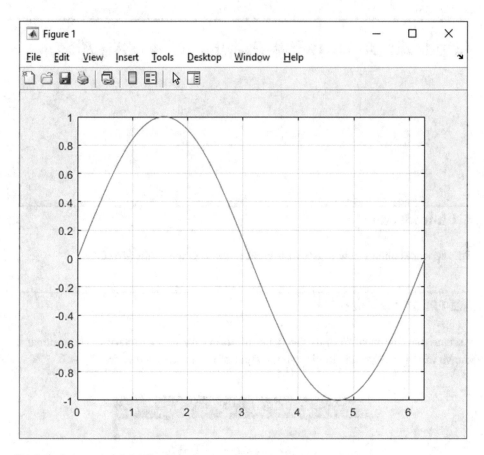

Fig. A.2 Output of code in Fig. A.1

You can click on any point in order to read its coordinate (Fig. A.3).

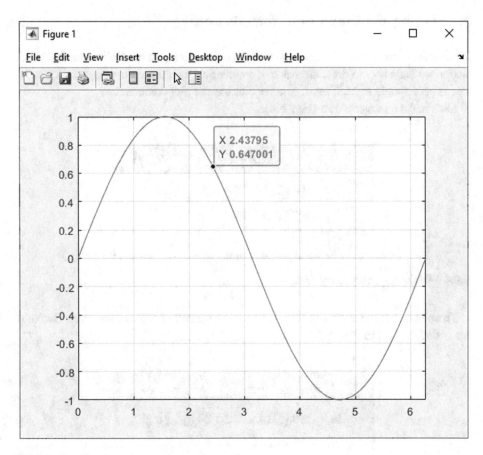

Fig. A.3 Reading the points on the graph

A.3 Plotting the Graph of a Numeric Data

In the previous section you learned how to draw the graph of a symbolic function. In this section we learn how to draw the graph of a numeric data. Plotting the graph of a numeric data is very simple in MATLAB. You need to use the plot command.

Let's make a numeric data (Fig. A.4).

```
Command Window                    ⊙
    >> x=[0:0.01:2*pi];
    >> y=sin(x);
fx >>
```

Fig. A.4 Making a simple sample data

The plot command shown in Fig. A.5 draws the graph of the numeric data. Output of this code is shown in Fig. A.6.

```
Command Window                    ⊙
    >> x=[0:0.01:2*pi];
    >> y=sin(x);
    >> plot(x,y)
fx >>
```

Fig. A.5 Drawing the graph of a numeric data with plot command

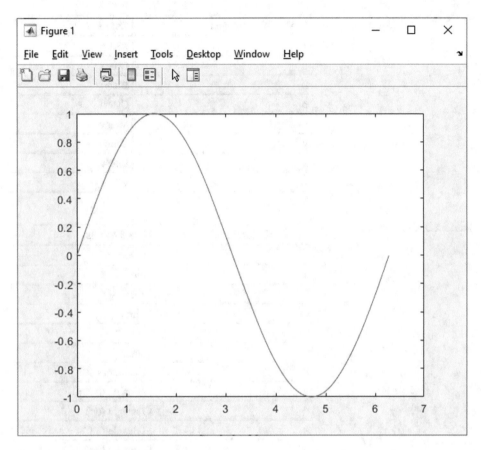

Fig. A.6 Output of code in Fig. A.5

You can add the operators shown in Tables A.1, A.2 and A.3 to produce more custom plots.

Table A.1 Types of lines

MATLAB command	Type of line
-	Solid
:	Dotted
-	Dashdot
-	Dashed

Table A.2 Colors

MATLAB command	Colors
r	Red
g	Green
b	Blue
c	Cyan
m	Magenta
y	Yellow
k	Black
w	White

Table A.3 Plot symbols

MATLAB command	Plot symbols
	Point
+	Plus
*	Star
O	Circle
x	x-mark
s	Square
d	Diamond
v	Triangle (down)
^	Triangle (up)
<	Triangle (left)
>	Triangle (right)

Let's study an example. Values of voltage and current for a resistor is shown in Table A.4. We want to plot the graph of this data. We want to show the data points with circles and connect them together using dashed line with black color. The vertical axis and horizontal axis must have the labels "Current (A)" and "Voltage (V)", respectively. The title of the graph must be "I–V for a resistor". The commands shown in Fig. A.7 do what we need. The result is shown in Fig. A.8.

Table A.4 V-I values for resistor R1

V (volt)	I (Amper)
0.499	0.10
0.985	0.20
1.508	0.31
1.969	0.41
2.528	0.53
2.935	0.61
3.481	0.73
3.971	0.83
4.486	0.94
4.960	1.04
5.502	1.15
6.007	1.26
6.60	1.38

```
Command Window                                                                    ⊙
  >> V=[0.499 0.985 1.508 1.969 2.528 2.935 3.481 3.971 4.486 4.960 5.502 6.007 6.60];
  >> I=[0.1 0.2 0.31 0.41 0.53 0.61 0.73 0.83 0.94 1.04 1.15 1.26 1.38];
  >> plot(V,I,'--ko')
  >> title('I-V for a resistor')
  >> xlabel('Voltage(V)')
  >> ylabel('Current(A)')
  >> grid on
fx >>
```

Fig. A.7 Drawing the graph of data in Table A.4

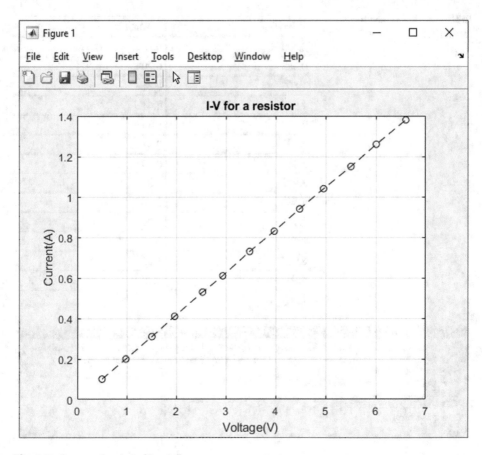

Fig. A.8 Output of code in Fig. A.7

The commands shown in Fig. A.9 draw the I–V graph of Table A.4. However, it uses red star for data points and solid black color for connecting the data points together (Fig. A.10).

```
Command Window
  >> V=[0.499 0.985 1.508 1.969 2.528 2.935 3.481 3.971 4.486 4.960 5.502 6.007 6.60];
  >> I=[0.1 0.2 0.31 0.41 0.53 0.61 0.73 0.83 0.94 1.04 1.15 1.26 1.38];
  >> plot(V,I,'k',V,I,'r*'),grid minor
fx >>
```

Fig. A.9 Drawing the graph of data in Table A.4

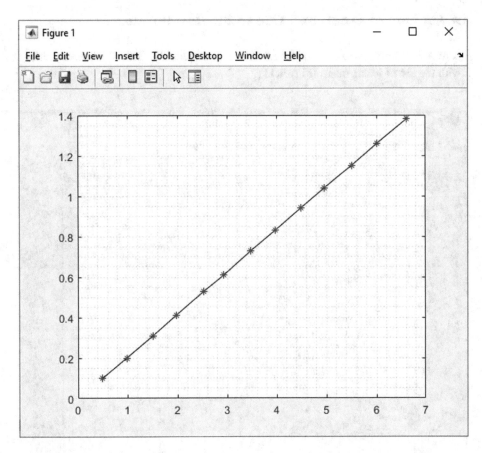

Fig. A.10 Output of code in Fig. A.9

A.4 Addition of Labels and Title to the Drawn Graph

Figure A.10 doesn't have any labels and title. You can add the desired labels and titles to it with the aid of insert menu (Fig. A.11)

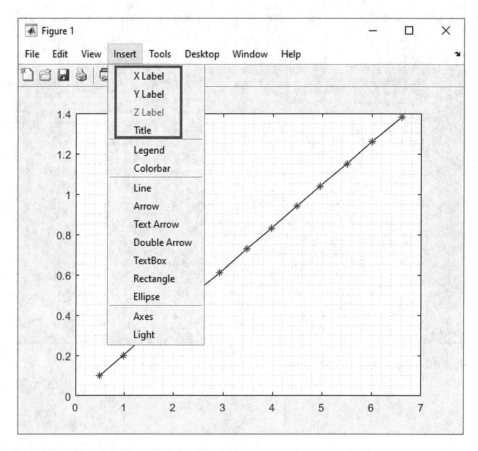

Fig. A.11 Addition of title and labels to the axis

A.5 Exporting the Drawn Graph as a Graphical File

You can copy the drawn graph to the clipboard easily with the aid of Edit > Copy Figure (Fig. A.12). After copying the graph to the clipboard you can easily paste it in programs like MS Word® by pressing Ctrl+V.

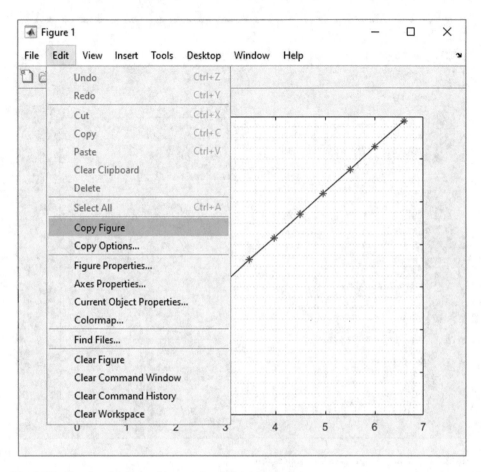

Fig. A.12 Copying the drawn figure to the clipboard memory

You can save the drawn graph as a graphical file as well. To do this, click use the File > Save As (Fig. A.13). After clicking, the save as window appears. Select the desired output format from the Save as type drop down list (Fig. A.14).

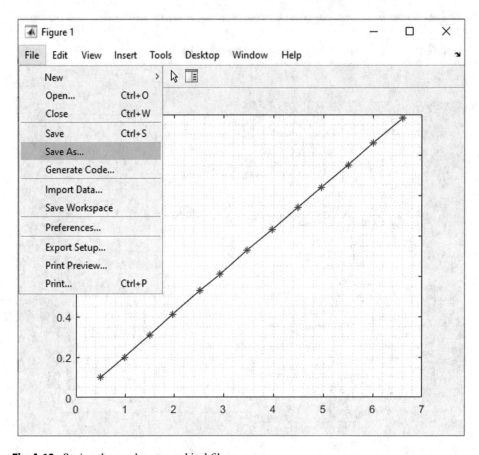

Fig. A.13 Saving the graph as a graphical file

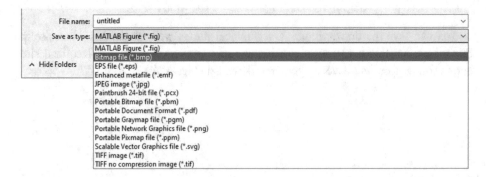

Fig. A.14 Selection of desired type of file

A.6 Drawing Two or More Graphs on the Same Axis

Sometimes you need to show two or more datasets on the same graph. You need to use the hold on command to show two or more graphs simultaneously. Assume that we have another dataset (Table A.5) and we want to show both datasets (Tables A.4 and A.5) on the same graph.

Table A.5 V-I values for resistor R2

V (volt)	I (Amper)
0.579	0.10
0.978	0.17
1.598	0.28
1.976	0.34
2.496	0.43
2.953	0.51
3.458	0.60
4.068	0.71
4.450	0.78
4.917	0.86
5.35	0.93
5.75	1.01
6.37	1.11
6.60	1.15

The commands shown in Fig. A.15 draws the graph of both datasets on the same graph. Output of this code is shown in Fig. A.16.

```
Command Window                                                              ⊙
 >> V1=[0.499 0.985 1.508 1.969 2.528 2.935 3.481 3.971 4.486 4.960 5.502 6.007 6.60];
 >> I1=[0.1 0.2 0.31 0.41 0.53 0.61 0.73 0.83 0.94 1.04 1.15 1.26 1.38];
 >> V2=[0.579 0.978 1.598 1.976 2.496 2.953 3.458 4.068 4.450 4.917 5.35 5.75 6.37 6.60];
 >> I2=[0.1 0.17 0.28 0.34 0.43 0.51 0.60 0.71 0.78 0.86 0.93 1.01 1.11 1.15];
 >> plot(V1,I1,'b',V1,I1,'r*')
 >> hold on
 >> plot(V2,I2,'k',V2,I2,'r+')
 >> grid minor
 >> xlabel('Voltage (V)')
 >> ylabel('Current (A)')
 >> title('Comparison of I-V plot of R1 and R2')
fx >>
```

Fig. A.15 Drawing the graph of Tables A.4 and A.5 on the same graph

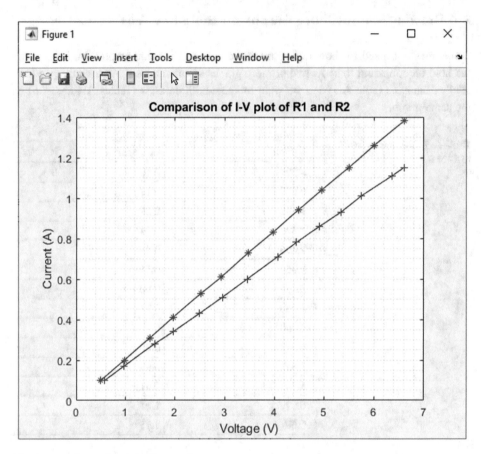

Fig. A.16 Output of code in Fig. A.15

You can use the Insert > Legend to show which graph belongs to which resistor (Fig. A.17).

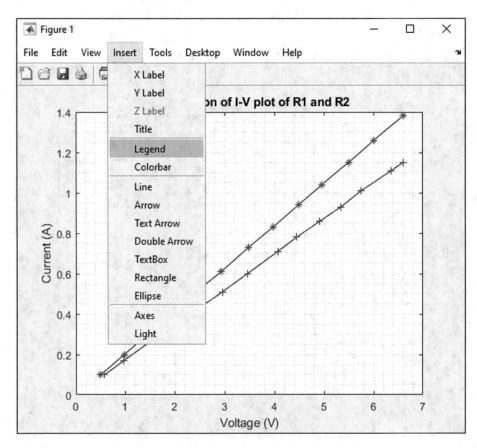

Fig. A.17 Insert > Legend can be used to add a legend to the graph

After clicking the Insert > Legend, the legend shown in Fig. A.18 will be added to the graph.

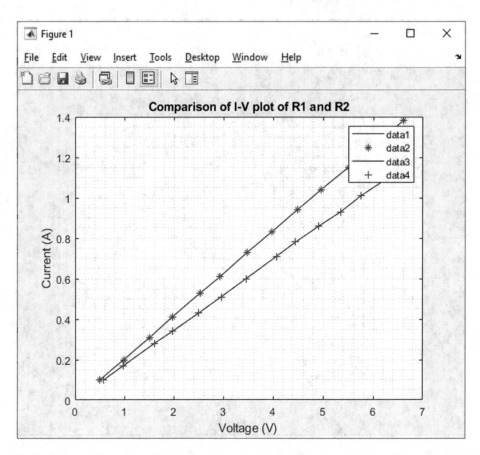

Fig. A.18 Legend is added to the graph

Double click the data1 in the legend box (Fig. A.18) and enter the desired text. Repeat this for data2, data3 and data4 in Fig. A.18. You can move the legend box by clicking on it, holding down the mouse button and dragging it to the desired location (Fig. A.19). You can even right click on the legend box and use the predefined locations (Fig. A.20).

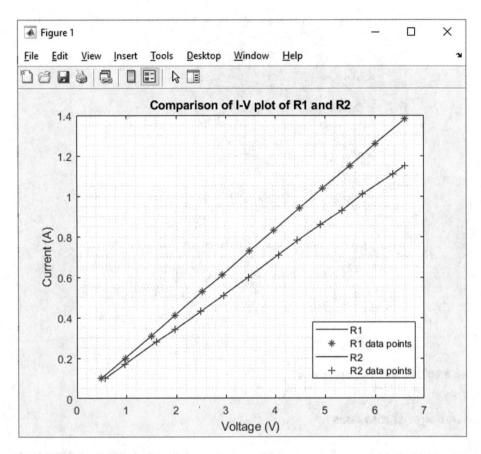

Fig. A.19 Customized legend

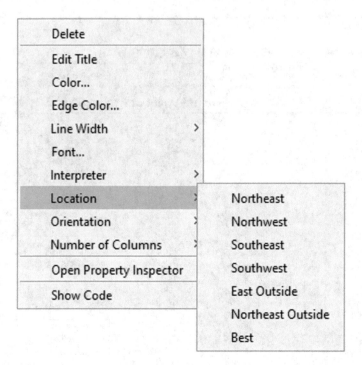

Fig. A.20 Default locations for legend

A.7 Logarithmic Axis

We used linear axis in order to draw the I–V graph of studied resistors. If you want to draw the frequency response graphs you need to use logarithmic axis. The linear axis is not a suitable option for frequency response graphs.

Let's study an example. Assume the frequency response given in Table A.6. This table shows the frequency response of the circuit shown in Fig. A.21.

Table A.6 Frequency response of a RC circuit

Frequency (Hz)	Magnitude $\left(\left\|\dfrac{V_o(j\omega)}{V_{in}(j\omega)}\right\|\right)$	Phase $\left(\left(\sphericalangle \dfrac{V_o(j\omega)}{V_{in}(j\omega)}\right)\right)$
1	1.000	−0.36°
10	0.998	−3.60°
20	0.992	−7.16°
50	0.954	−17.44°
100	0.847	−32.13°
150	0.728	−43.30°
200	0.623	−51.48°
250	0.537	−57.51°
300	0.469	−62.05°
350	0.414	−65.54°
400	0.370	−68.30°
450	0.333	−70.51°
500	0.303	−72.34°
550	0.278	−73.85°
600	0.256	−75.14°

Fig. A.21 Simple RC circuit

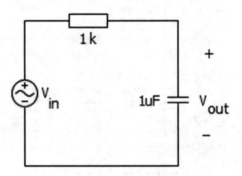

The commands shown in Fig. A.22, draws the frequency response of the data in Table A.6. The command semilogx is used to draw the frequency response graph. Output of this code is shown in Fig. A.23.

```
Command Window
  >> f=[1 10 20 50 100 150 200 250 300 350 400 450 500 550 600];
  >> Amp=[1 0.998 0.992 0.954 0.847 0.728 0.623 0.537 0.469 0.414 0.370 0.333 0.303 0.278 0.256];
  >> Phase=-[0.36 3.6 7.16 17.44 32.13 43.30 51.48 57.51 62.05 65.54 68.30 70.51 72.34 73.85 75.14];
  >> subplot(211),semilogx(f,20*log10(Amp)),grid minor
  >> title('Frequency responce of the RC circuit')
  >> xlabel('Freq(Hz.)')
  >> ylabel('Amplitude (dB)')
  >> subplot(212),semilogx(f,Phase),grid minor
  >> xlabel('Freq(Hz.)')
  >> ylabel('Phase(Degrees)')
fx >> |
```

Fig. A.22 Drawing the graph of data in Table A.6

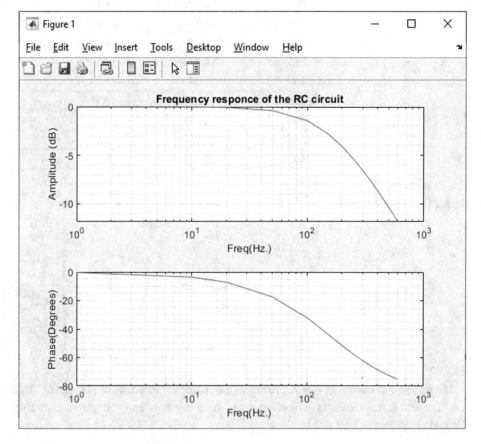

Fig. A.23 Output of code in Fig. A.22

Appendix B: Circuit Design with Circuit Wizard

B.1 Introduction

Multisim™ has a powerful tool that permits you to design different types of circuits quickly and easily. In this chapter you will learn how to use this tool to design what you want.

B.2 555 Timer Wizard

You can use the Tools > Circuit wizard > 555 timer wizard (Fig. B.1) to design astable and monostabe circuits based on the 555 timer IC. After clicking the 555 timer wizard, the window shown in Fig. B.2 appears.

© The Editor(s) (if applicable) and The Author(s), under exclusive license   197
to Springer Nature Switzerland AG 2023
F. Asadi, *Analog Electronic Circuits Laboratory Manual*, Synthesis Lectures on Electrical
Engineering, https://doi.org/10.1007/978-3-031-25122-1

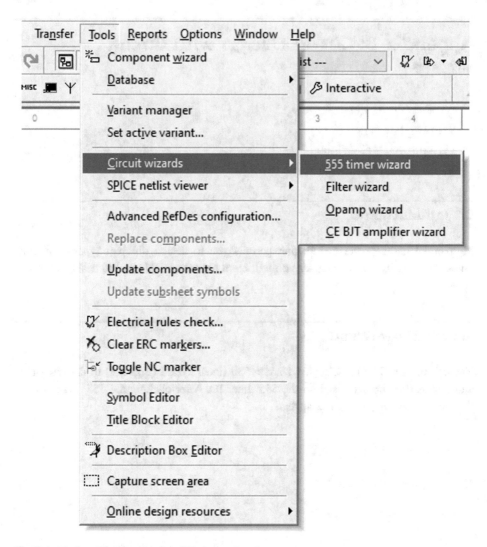

Fig. B.1 Tools > Circuit wizards > 555 timer wizard

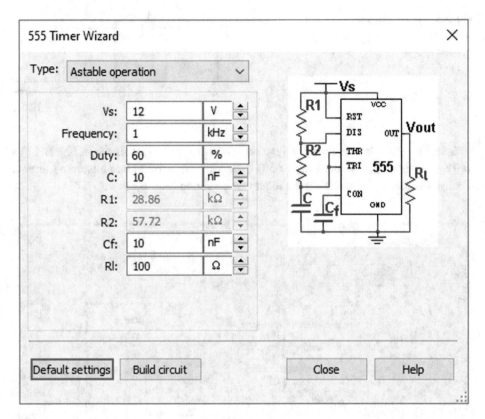

Fig. B.2 555 timer wizard

Assume that we want to generate a square wave similar to what shown in Fig. B.3 with $A = 12\,\text{V}$, $T = 1\,\text{ms}$ and $f = \frac{1}{T} = 1\,\text{kHz}$. Duty cycle is defined as the ratio of duration of High portion of the pulse to the period. For instance, in Fig. B.3, duration of High portion is T/2. Ratio of High portion to the period is T/2/T = 0.5 or 50%.

Fig. B.3 Typical square
waveform

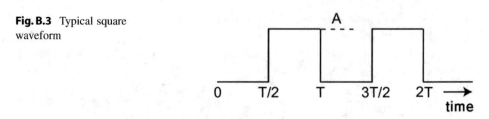

Select the Astable operation for Type box and enter the values shown in Fig. B.4 (Vs = 12, Frequency = 1 k and Duty = 50). Note that the wizard gives a warning to you. Change the value of C to satisfy the requested condition (Fig. B.5).

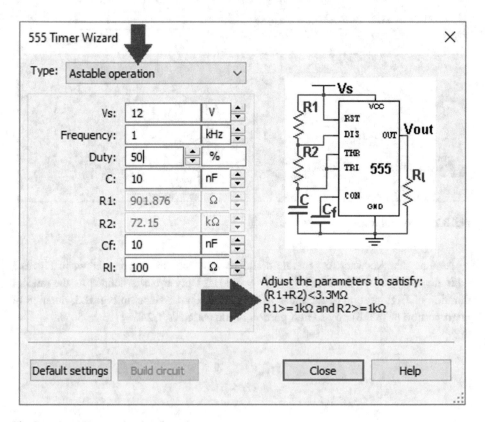

Fig. B.4 Astable operation is selected

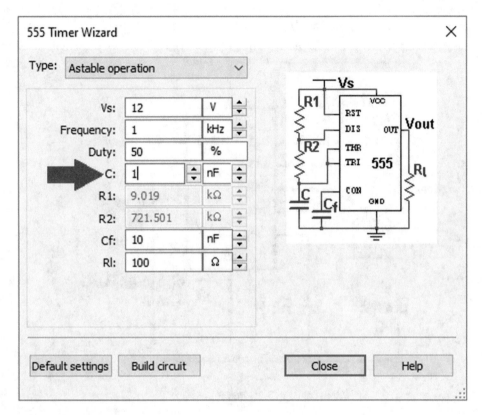

Fig. B.5 Value of capacitor C is decreased to 1 nF

Click the Build circuit button in Fig. B.5 in order to add the designed circuit to the schematic (Fig. B.6).

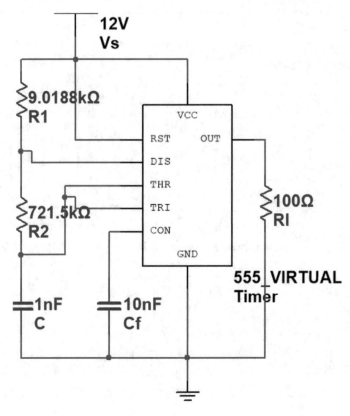

Fig. B.6 Designed circuit

Let's see whether the circuit works as expected. Add an oscilloscope to the circuit output (Fig. B.7) and run the simulation. Result is shown in Fig. B.8. According to Fig. B.8 frequency of the generated waveform is around $\frac{1}{1}.013\,m = 987.1668\,Hz$ which is quite close to 1 kHz. You can measure the duty ratio and amplitude with the aid of cursors and ensure that they are quite close to the expected values.

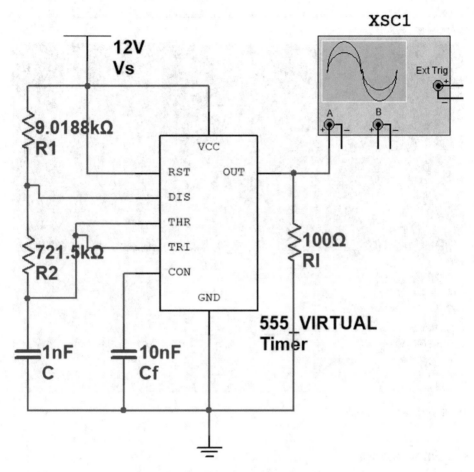

Fig. B.7 An oscilloscope is connected to the output of the circuit

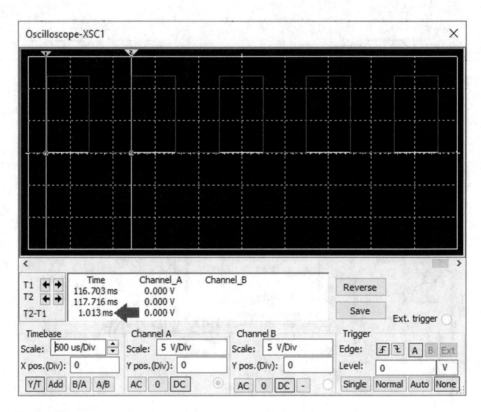

Fig. B.8 Simulation result

B.3 Filter Wizard

You can use the Tools > Circuit wizard > Filter wizard to design different types of filters (i.e., low pass, high pass, band pass and band reject). After clicking the Tools > Circuit wizard > Filter wizard, the window shown in Fig. B.9 appears on the screen.

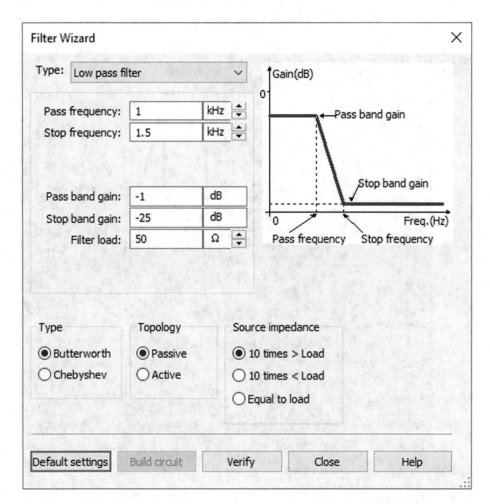

Fig. B.9 Filter wizard window

Use the Type drop down list to select the desired type of filter (Fig. B.10). The Filter load box in Fig. B.10 determines the output load of the filter (Load resistor shown in Fig. B.11). The Topology section determines whether the filter is realized with active (Op Amp) or passive (R, L and C) components. The picture shown in Fig. B.10 clearly defines the pass band gain, pass frequency, stop frequency, …

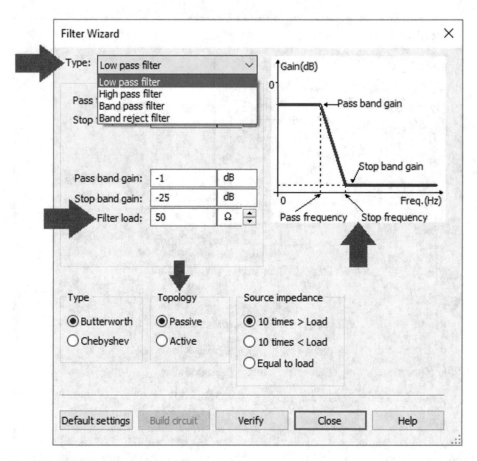

Fig. B.10 Different parts of filter wizard

Fig. B.11 Value of load
resistor is determined by the
filter load box in Fig. B.10

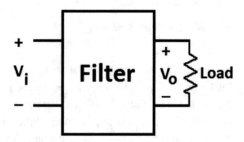

Let's design a low pass active Butterworth filter. Butterworth filters are well known for their peak free frequency response. Filter load box in Fig. B.12 is filled with 100 Ω therefore output load (Load resistor shown in Fig. B.11) equals to 100 Ω. In Fig. B.12, Resistance in LP box is filled with 1000 Ω. So, 1000 Ω resistors will be used to design the filter block.

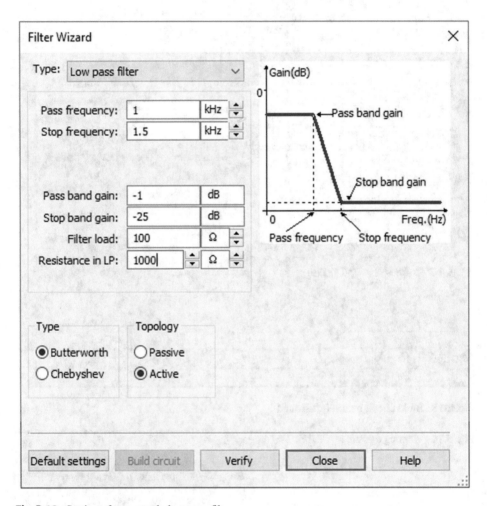

Fig. B.12 Settings for a sample low pass filter

Click the verify button in Fig. B.B. When your design is error-free, a "Calculation successful" message displays and Build circuit button becomes active (Fig. B.13). Click the Build circuit button. After clicking the Build circuit button, the warning shown in Fig. B.14 may appear. Click the Yes button to continue. After clicking the Yes button, click on the schematic to paste the designed circuit to it (Fig. B.15).

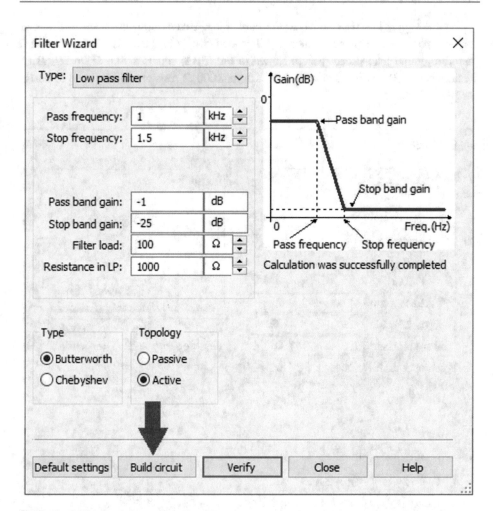

Fig. B.13 Build circuit button is activated

Fig. B.14 Warning message

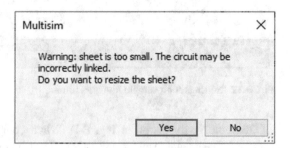

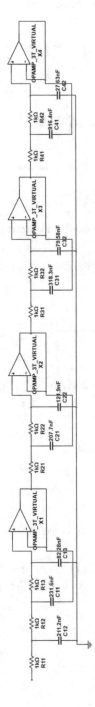

Fig. B.15 Designed circuit

Add an AC voltage source to the input of the filter (Fig. B.16).

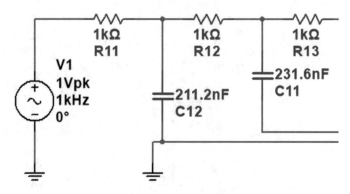

Fig. B.16 Voltage source V1 is connected to input of the circuit

Add a 100 Ω resistor to the output of the filter and give the name output to the output node (Fig. B.17).

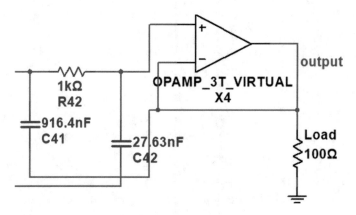

Fig. B.17 100 Ω load resistor is connected to the output of the circuit

Now we can use the AC sweep in order to ensure that frequency response of designed circuit is what we need. Click the Interactive button (Fig. B.18) and set up an AC sweep with settings shown in Figs. B.19 and B.20.

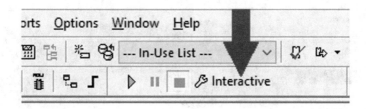

Fig. B.18 Interactive button

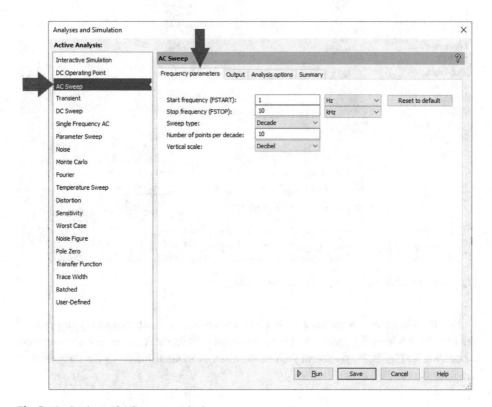

Fig. B.19 Settings of AC sweep analysis

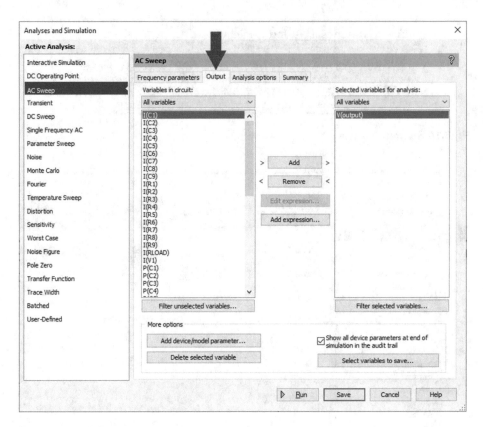

Fig. B.20 Determining the output of AC sweep analysis

Run the simulation by clicking on the Run button in Fig. B.20. Frequency response of the filter is shown in Fig. B.21. Now you can decide whether this circuit is what you need. According to Fig. B.22, passband gain of the filter is around 0 and -3 dB frequency of the filter is around 1 kHz.

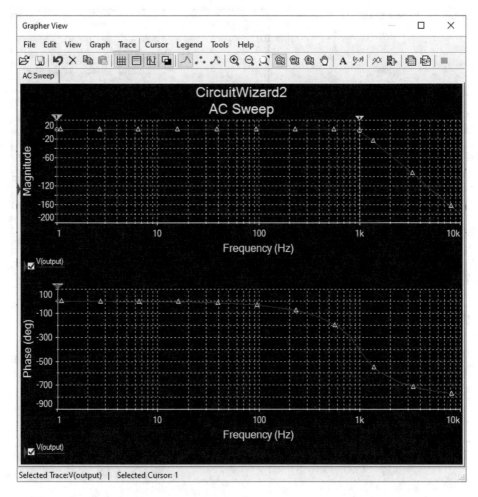

Fig. B.21 Simulation result

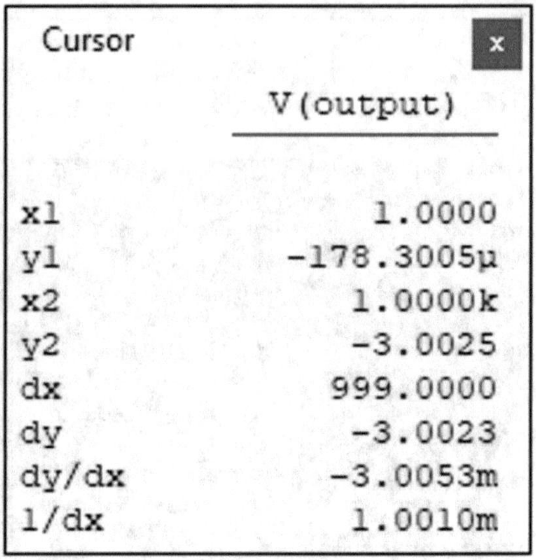

Fig. B.22 Cursor window

Cursor		x
	V(output)	
x1	1.0000	
y1	−178.3005μ	
x2	1.0000k	
y2	−3.0025	
dx	999.0000	
dy	−3.0023	
dy/dx	−3.0053m	
1/dx	1.0010m	

B.4 Opamp Wizard

You can use the Tools >Circuit wizard >Opamp wizard to design different types of Op Amp amplifiers. Opamp wizard is shown in Fig. B.23.

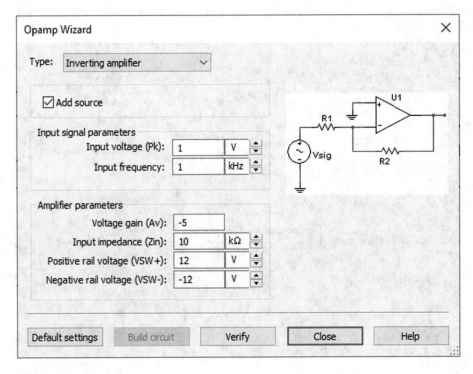

Fig. B.23 Opamp wizard

Use the Type drop down list to select the type of amplifier that you want (Fig. B.24).

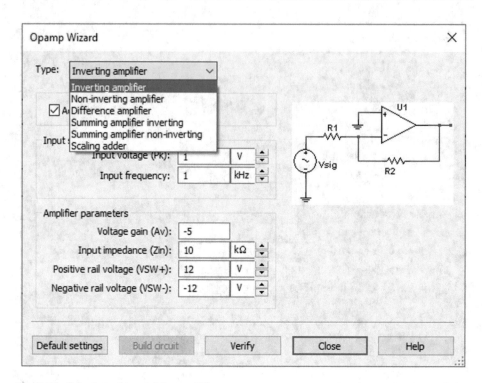

Fig. B.24 Selection of type of the amplifier

For instance, assume that we want to design a non-inverting summing amplifier with 3 inputs and input-output relation of $v_{out} = 20v_{ave} = 20 \times (\frac{v_1+v_2+v_3}{3})$. Settings shown in Fig. B.25 make such an input-output relation. After entering the settings, click the verify button (Fig. B.25). Then click the Build circuit button (Fig. B.25) to design the circuit. Designed circuit is shown in Fig. B.26. Note that values of feedback resistor Rf (resistor between output and negative terminal of the Op Amp), R1, R2 and R3 is determined by the value entered to the Feedback resistor value (Rf) box in Fig. B.25.

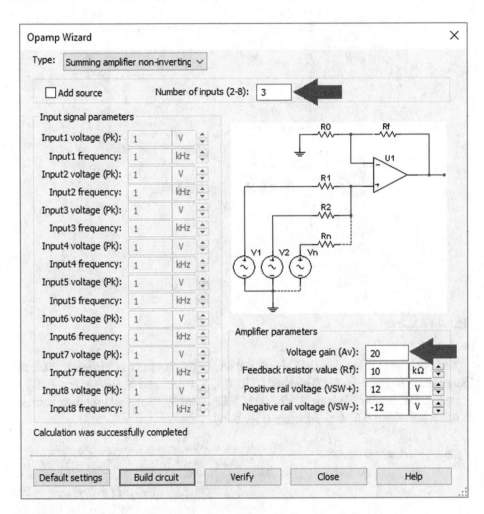

Fig. B.25 Design of non-inverting summer amplifier with three inputs

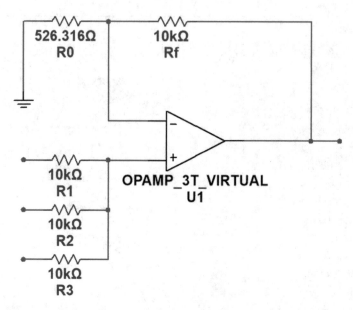

Fig. B.26 Designed circuit

Let's give input to the designed circuit and ensure that it works as expected. Click the Place Source button (Fig. B.27) and add V_REF 1, V_REF 2 and V_REF 3 (Fig. B.28) to the schematic. Double click on V_REF 1, V_REF 2 and V_REF 3 and set their voltages to 0.2 V, 0.4 V and 0.6 V, respectively (Fig. B.29).

Fig. B.27 Place source button

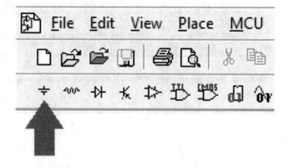

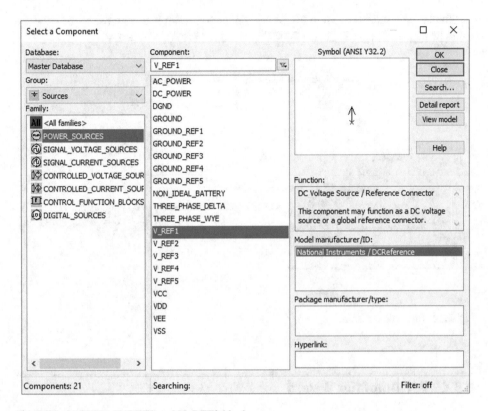

Fig. B.28 V_REF1, V_REF2 and V_REF3 blocks

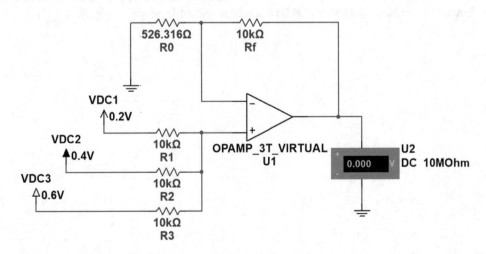

Fig. B.29 V_REF1, V_REF2 and V_REF3 blocks are connected to the inputs

Simulation result is shown in Fig. B.30. Simulation result is quite close to the expected value of $v_{out} = 20v_{ave} = 20 \times \left(\frac{v_1+v_2+v_3}{3}\right) = 20 \times \left(\frac{0.2+0.4+0.6}{3}\right) = 8$ V.

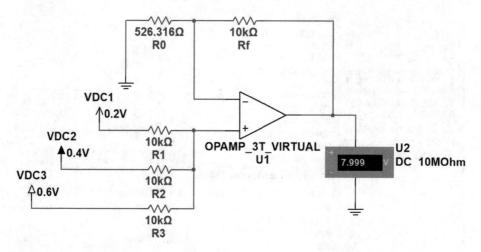

Fig. B.30 Simulation result

B.5 CE BJT Amplifier Wizard

You can use the Tools > Circuit wizard > CE BJT amplifier wizard to design Common Emitter single stage amplifiers. CE BJT amplifier wizard is shown in Fig. B.31.

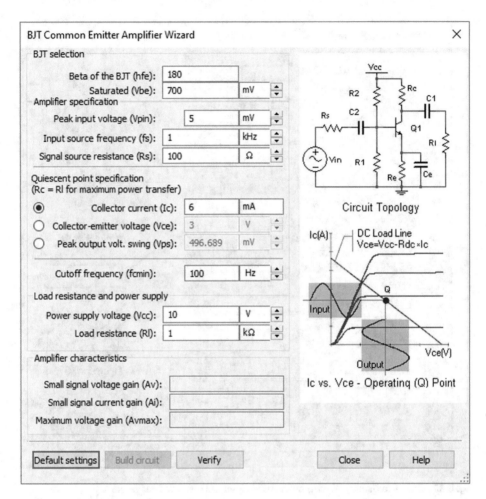

Fig. B.31 BJT common emitter amplifier wizard

Let's design an amplifier with CE BJT amplifier wizard. Assume that we need an amplifier with lower cut-off frequency of 20 Hz. Enter 20 to the Cutoff frequency (fcmin) box (Fig. B.32).

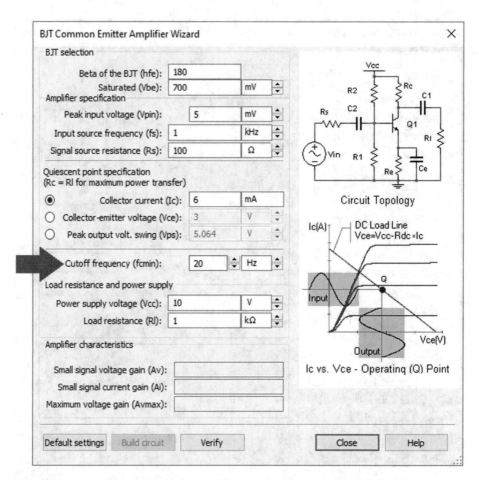

Fig. B.32 Cutoff frequency (fcmin) box is used to set the lower cut-off frequency

Let's assume our supply voltage is 12 V. Enter 12 to the Power supply voltage (Vcc) box (Fig. B.33). Let's keep other values unchanged.

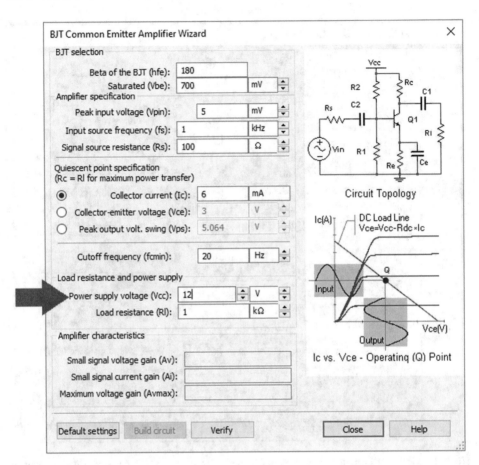

Fig. B.33 Power supply voltage (Vcc) box is used to set the value of power supply (VCC)

Click the Verify button in Fig. B.33. For Collector current (Ic) of 6 mA, the small signal voltage gain is 101.275319 V/V (Fig. B.34). You can obtain the desired voltage gain by changing the collector current. Note that collector current is directly related to the amplifier voltage gain: Higher collector current means bigger voltage gain and lower collector current means smaller voltage gain. For instance, if you decrease the collector current to 5 mA, the small signal voltage gain decreases to 86.390785 V/V (Fig. B.35). If you increase the collector current to 7 mA, the small signal voltage gain increases to 114.517693 V/V (Fig. B.36).

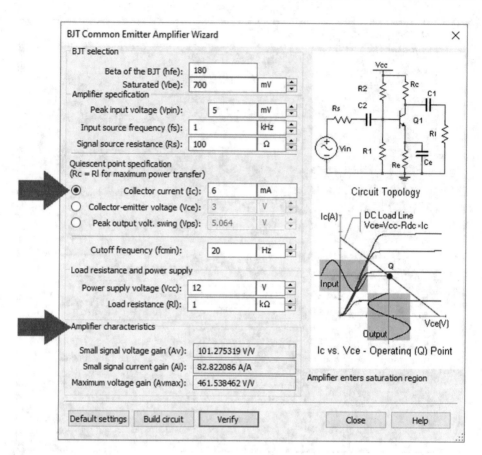

Fig. B.34 Collector current (Ic) is used to set the collector current. Small signal voltage gain is around 101.27 for collector current of 6 mA

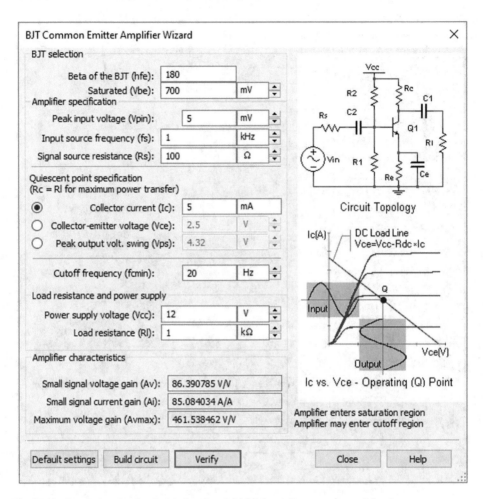

Fig. B.35 Small signal voltage gain is around 86.39 for collector current of 5 mA

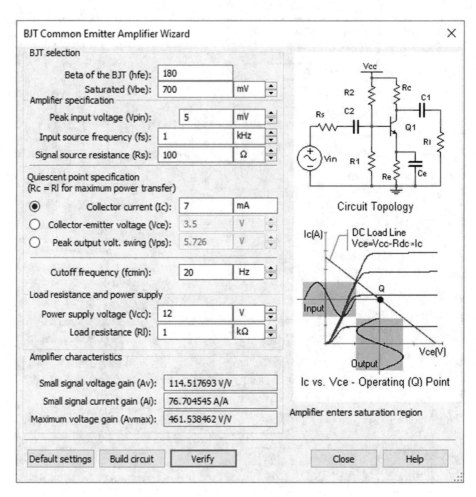

Fig. B.36 Small signal voltage gain is around 114.52 for collector current of 7 mA

You can build the circuit by clicking the Build circuit button. For instance, Fig. B.37 shows the designed circuit for collector current of 6 mA.

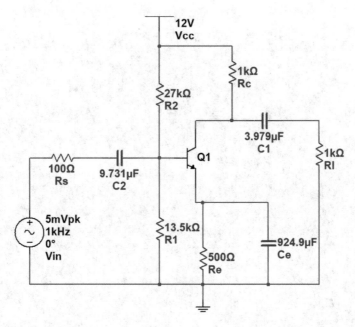

Fig. B.37 Designed circuit

Figure B.38 shows the frequency response of designed circuit. Note that magnitude graph is flat even up to 10 MHz. Let's change the transistor and see what happens. Double click on transistor Q1 and click the Replace button

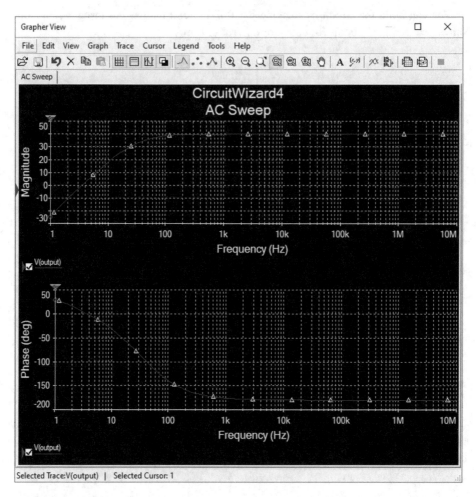

Fig. B.38 Simulation result

Let's use the cursors to measure the low frequency -3 dB cut-off frequency. According to Figs. B.39 and B.40, the mid-band voltage gain is 39.8268 dB which equals to gain of $10^{\frac{39.8268}{20}} = 98.0257\frac{V}{V}$. Low frequency -3 dB cut-off frequency is around 59.4762 Hz. So, the cut-off frequency is bigger than what we expected. Increase the C1, C2 and Ce by factor of $\frac{59.4762}{20} \approx 3$. Figure B.41 shows the frequency response of the circuit for new values of capacitors. According to Fig. B.42, the low frequency -3 dB cut-off frequency is around 20 Hz. Note that mid-band gain is not affected from this capacitor value changes.

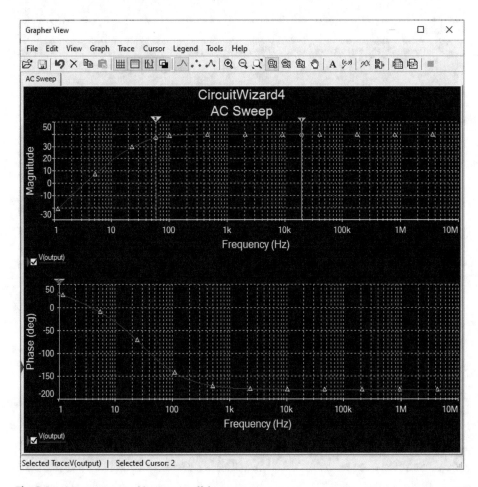

Fig. B.39 Measurement of lower cut-off frequency

Fig. B.40 Cursor window

Cursor	
	V(output)
x1	19.7337k
y1	39.8268
x2	59.4762
y2	36.8079
dx	-19.6742k
dy	-3.0188
dy/dx	153.4400μ
1/dx	-50.8279μ

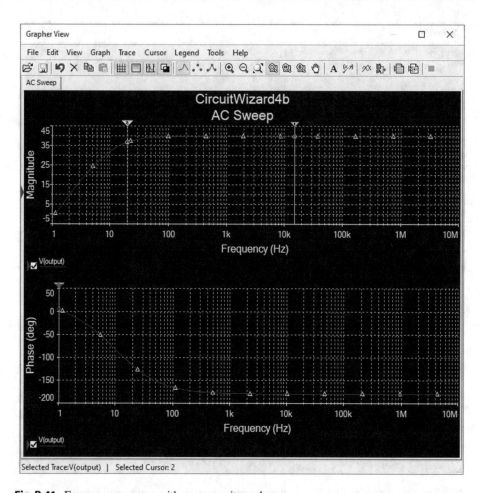

Fig. B.41 Frequency response with new capacitor values

Fig. B.42 Cursor window

Cursor	x
	V(output)
x1	15.2334k
y1	39.8268
x2	20.0021
y2	36.8783
dx	-15.2134k
dy	-2.9485
dy/dx	193.8106µ
1/dx	-65.7317µ

Let's replace the transistor in Fig. B.37 with a 2N2222 and see what happens to the frequency response. Double click the transistor in Fig. B.37. This opens the BJT_NPN window (Fig. B.43).

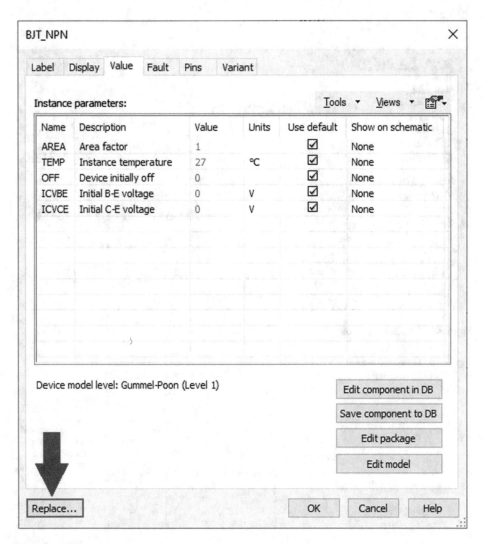

Fig. B.43 BJT_NPN window

Click the Replace button in Fig. B.43 and after that select the 2N2222 (Fig. B.44). Now the schematic changes to what shown in Fig. B.45.

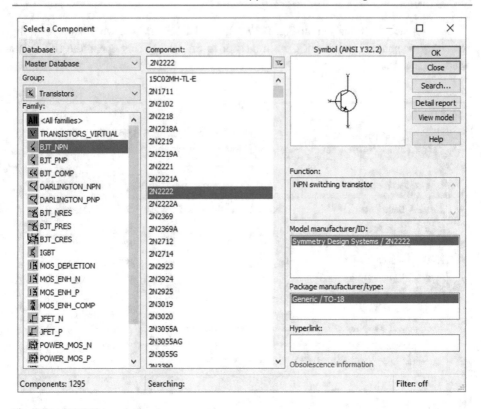

Fig. B.44 2N2222 transistor

Fig. B.45 Designed circuit
with 2N2222 transistor

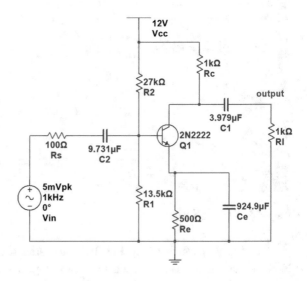

Frequency response of Fig. B.45 is shown in Fig. B.46. Note that the magnitude graph starts to fall around 1 MHz.

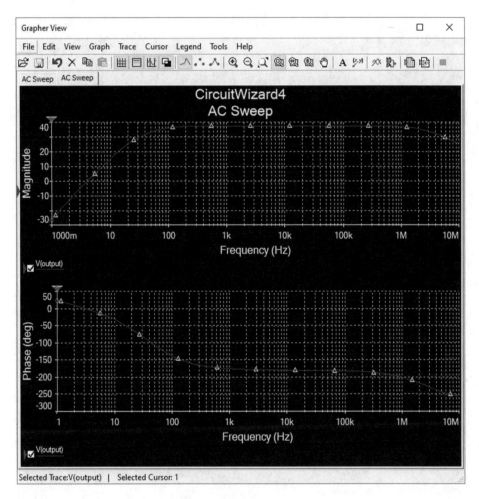

Fig. B.46 Frequency response with 2N2222 transistor

Printed in the United States
by Baker & Taylor Publisher Services